Fabien Gouyon

Computational Rhythm Description

Fabien Gouyon

Computational Rhythm Description

A Review and Novel Approach

VDM Verlag Dr. Müller

Imprint

Bibliographic information by the German National Library: The German National Library lists this publication at the German National Bibliography; detailed bibliographic information is available on the Internet at http://dnb.d-nb.de.

Cover image: www.purestockx.com

Publisher:
VDM Verlag Dr. Müller Aktiengesellschaft & Co. KG, Dudweiler Landstr. 125 a, 66123 Saarbrücken, Germany,
Phone +49 681 9100-698, Fax +49 681 9100-988,
Email: info@vdm-verlag.de

Produced in USA and UK by:
Lightning Source Inc., La Vergne, Tennessee, USA
Lightning Source UK Ltd., Milton Keynes, UK
BookSurge LLC, 5341 Dorchester Road, Suite 16, North Charleston, SC 29418, USA

ISBN: 978-3-8364-7769-7

Acknowledgments

It goes without saying that the work reported in this document is not the work of a single person. It really is a collaborative effort and, at the risk of unfair omission, I want to express my gratitude to those who participated.

I have been extremely lucky to reside for five years in the Music Technology Group of Pompeu Fabra University in Barcelona, where I could interact with such brilliant and insightful researchers. The first person I would like to thank is Xavier Serra for inviting me to join the Music Technology Group, when we met at CCRMA back in 1999, and for encouraging and supporting me since then. I am also grateful for the multiple opportunities I had to make my work public. I truly enjoyed working with many people in the MTG and I want to thank them here for the interesting discussion we had and for proof-reading writings of mine: Àlex Loscos, Martin Kaltenbrunner, Günter Geiger, Álvaro Barbosa, Lars Fabig, Emilia Gómez, Jordi Bonada, Jordi Janer, Sergi Jordà, Ross Bencina, Bram de Jong, Sebastian Streich, Bee-Suan Ong, Oscar Celma, Markus Koppenberger, José-Pedro García, Hendrik Purwins, Nicolas Wack, Oscar Mayor, Hugo Solís, Amaury Hazan and Maarten Grachten. I also want to thank Marteen de Boer, Ramón Loureiro and Carlos Atance for providing the means to work in amazingly smooth material conditions. Other technical thanks go to David García and Miquel Ramírez. Thanks also to Joana Clotet and Cristina Garrido for helping in all administrative tasks and, most importantly, for creating such a nice atmosphere around them.

Pedro Cano, Perfecto Herrera and Simon Dixon deserve a special mention for their enormous (and I mean it) influence on my work, for being amazing sources of knowledge and inspiration and always being willing to spend a couple of hours (with

2

or without beer or "vieille prune") discussing this or that aspect of some scientific problem, or proof-reading hundreds of pages. Special credits should also be given to Xavier Serra and Gerhard Widmer, my supervisors, for their expert guidance throughout the years.

I also wish to reflect on the past and thank the people who brought me to this point (writing the acknowledgments of a book). The first credits should be given to my parents, Jean-Paul and Christine Gouyon, my elder sister Valérie Brunot and my grand-parents, Georges and Jeanne Gouyon and Lucienne Bonin, for their love and support of course but also for transmitting to me the love of learning. I also thank Zizou and Daniel Tardivon for their unconditional support while I lived in Paris as well as Pierre-François, Lise and Fanny Brunot for their love.

Many other people (professors, work colleagues or just fellow researchers with whom I had the chance to have insightful discussions) influenced my research in the last years. I want to thank them here, in a chronological order: Régine André-Obrecht, Philippe Lepain and Philippe Joly (at IRIT), Francis Castanié, Jean-Yves Tourneret and Corinne Mailhes (at ENSEEIHT), Julius Orion Smith III, Harvey Thornburg, Stefan Bilbao, Bob Sturm, Hendrik Purwins, Tamara Smyth, Stefania Serafin, Perry Cook, Juan Pampin, David Berners, Fernando Lopez-Lezcano, Dan Levitin and Caroline Traube (at CCRMA), Jeremy Marozeau, Thibaut Ehrette, Emilia Gómez, Xavier Rodet, Olivier Warusfel, Jean-Claude Risset, Gérard Assayag, Daniel Arfib, Richard Kronland-Martinet, Philippe Depalle, Stephen McAdams, Alain de Cheveigné, Hughes Vinet, Geoffroy Peeters, Olivier Lartillot and Benoit Meudic (at IRCAM), François Pachet, Olivier Delerue, Luc Steels and Gert Westermann (at Sony CSL), Elias Pampalk and Werner Goebl (at ÖFAI), Anssi Klapuri (with a special thanks for his thorough and inspiring review of this document), Mark Sandler, Juan-Pablo Bello, George Tzanetakis, Tristan Jehan, Daniel Pressnitzer, Peter Desain, Martin McKinney, Douglas Eck, Robert Zatorre, Barbara Tillmann, Sofia Dahl and Guy Madison. More closely related to this document, I also want to thank the coauthors of my papers who all contributed an important part of the work presented here and anonymous reviewers of these papers who (usually) helped in improving them.

For being great friends, making Barcelona a place where it is so difficult to work and (for some of them) being able to cope with my moody behavior in the last months of writing my thesis, I would like to thank Pedro Cano, Chiara Capuccio, Martin Kaltenbrunner, Lars Fabig, Jordi Janer, Marie-Florence Deruffi, Daniele Bertolucci, Günter Geiger, Nadine Alber, Álvaro Barbosa, Cristina Alves de Sá, Joachim and Line Haas, Àlex Loscos, José-Pedro García, Markus Koppenberger, Rossella Cascone, Marion Hassler, Alejandro Ramirez, Cecilia Gayon, Vadim Tarasov, Mila Makarevitch, Emilia Gómez, Jordi Bonada, Maira Laranjeira, Aurora Iraita, Laura Colloridi, Begonya Samit, Hagar Zeligman, Arantxa, Silvia Poblete, Claudia Rivera, Macarena Francia, los del cajón, de "la Fuga" and Joana Martins. But Barcelona is not all... In other parts of the world, other people cannot go unmentioned for they shared fun parts (and helped me through hard parts) of my life since I began this work. I wish to thank them in a geographical order (why not?, from the closest to the farthest): Ivan and Colin Segalowitch, Philippe Nadal, Philippe Terral, Jordan Treffé, Christine and José Moraté, Marianne Chambon, Camille Bonaldi, Alice and Étienne Bertaud du Chazaud, Béatrice Boué, Michael Janus, Gerda Strobl, Claudia Kogler, Eva Kahler, Harald Mayer, Claudia Vargas and Katsuhiko Sakamoto.

Energy was supplied by 30 plats, el Salvador, el Rodri, el Sarima, el Foro, el Económic, "el de en frente del Económic," el Rincón de la Ciutadela, el Pasatore, "los pescaditos," la Plata, el Refugio del Puerto and of course... "las alemanas." Burning energy excesses was possible thanks to the Club Natació Atlètic-Barceloneta and the beach on which I could run while admiring such wonderful scenes.

4

Contents

Chapter 1

Introduction

Musical signals convey rhythm and when listening to music, or even barely hearing it, people experience rhythm. People without musical training (other than having been exposed to music since childhood) as well as trained musicians (trained to play Jazz, Pop or Baroque music for instance) hear rhythm in musical audio signals. Activities as whistling on the way to the office, tapping or swinging along some music in the background, synchronizing with other musicians while playing, dancing, or simply enjoying a piece of music (in fact, any activity linked to music) involve a *sense of rhythm* that seems to be constantly encountered in human beings.

Computers, on the other hand, have no sense of rhythm. Involving a machine in music (in any of the musical activities mentioned above) requires the implementation of software or hardware components that explicitly handle rhythm. Yet, in spite of the fact that rhythm is ubiquitous in our lives, we still do not possess a complete understanding of what it really is and which are the processes involved in our perception of it.

In this book, we address the definition of rhythmic representations of audio signals which, we argue, make up a central element in the handling of musical rhythm by computer systems. We call these representations *rhythm periodicity functions*. This book describes and evaluates several strategies to compute rhythm periodicity functions and demonstrates the usefulness of such functions in several situations where the computer is asked to make human-like musical responses to music as for instance "perceiving" tempo and beats, estimating rhythmic similarity or recognizing diverse musical genres.

1.1 Rhythm periodicity functions

Rhythm is about recurring musical events occurring roughly periodically. We do not yet know what "musical events" are to our perceptual system, how similar two events must be to be categorized as *the same* event, nor how we actually measure their recurrences, which are acceptable deviations for perfect periodicity and which processes further lead to the perception of rhythm. A cornerstone of this book lies in the assumption that the computational modeling of rhythm perception can be broken down into several functional modules. We argue that the first such module consists in the parsing of musical signals into low-level continuous temporal features emphasizing rhythmic aspects. The second functional module then seeks feature periodicity in time. We argue that a convenient way to do this is to generate continuous functions representing the salience of feature periodicity versus period (or inversely frequency): rhythm periodicity functions.

1.2 Methodology

Rhythmic aspects of music have been studied for many years, within many disciplines and by means of different methodologies, among which the following ones:

Theoretical: Theoretical analyses of music formulate theories about the syntactic structure of music and ways to notate it. Rhythmic aspects have been widely studied via this research paradigm (Cooper and Meyer, 1960; Yeston, 1976; London, 2005).

Behavioral: Some researchers point out that rhythm should not refer to properties of the musical notation system but rather to the experience aspects of *listeners*. Music psychology methodologies are usually empirical: a hypothesis is formulated, stimuli are generated in accordance with this hypothesis (usually a single dimension of interest is carefully controlled), listeners are selected, a procedure is defined (apparatus, indications to listeners, etc.), data is collected and finally analyzed in order to accept or reject the hypothesis. See e.g. (Fraisse, 1982) and (Desain and Windsor, 2000, Part III).

The *production* of rhythmic sequences, and their physiological and cognitive aspects, are also usually studied via experimental settings (e.g. synchronization or continuation tasks). Timing deviations in rhythm production are of first interest here (Madison, 2000), especially intentional expressive deviations produced by music performers (Palmer, 1997; Clarke, 1999). The study of performer gestures in the production of rhythmic sequences, and the associated communication of emotional features, also often implies empirical studies (Dahl, 2005).

Computational: Since the beginning of the computer age, people have been trying to build systems that could analyze and perform music. Along this engineering line of research, an important goal is the design of systems that can perform automatically, and reliably, a task normally suited to humans. Many references of computational approaches to rhythm understanding will be given in this rest of this book.

Neurophysiological: In the neurophysiological approach, the aim is the analysis of the brain anatomical structures involved in auditory processes (Zatorre, 2005). This entails issues of localization of such structures (often tackled by brain imaging methods (Levitin and Menon, 2005)) as well as observation of temporal coordinations between different functional modules (Desain, 2004).

The computational approach to neuroscience (the design of artificial models of neurons and neural networks) provides additional insights in the understanding of cerebral structures and processes involved in audition (Husain et al., 2004), (Tillmann et al., in press).

Our approach is only computational. We acknowledge that rhythm is a perceptual entity, yet this book does not propose a theory of rhythm perception, nor a principled model of auditory processes involved in rhythm perception. We argue however that computational models should not necessarily be seen as posterior implementations of established perceptual theories. Designing machines that can perform certain tasks "similarly" to humans has always been a motivation to science, even if the processes implemented are not grounded perceptually or cognitively; and engineering working systems often leads to the development of hypotheses regarding human capabilities. As Scheirer argues, "computational models and psychoacoustic experiments play overlapping and complementary roles in advancing our knowledge about the world" (2000, p.66).

On the other hand, a key point in a computational approach lies in the measure of this "similarity" between machine and human behaviors. In this book, this similarity is defined directly by the difference between algorithm outputs and reference annotations put manually by humans (the "ground-truth"). This is arguable, indeed, the definition of a methodology, metrics and even ground-truth data for the evaluation of computational models of music cognition in general (and automatic rhythm description in particular) is currently an active (and controversial) topic.[1] Nevertheless, it is our belief that the design of computational models must not necessarily wait for a consensus on the way to evaluate them (and thus wait for a complete understanding of the very problem they address). On the contrary, as we already argued in (Gouyon et al., 2006), the design of a computational model is often a way to better define the very problem under study; and the requirement of a sound evaluation of the model enforces the emergence of an agreement on the manner of representing and annotating relevant information about musical data, reference (publicly-available) examples of correct analyses and agreed evaluation metrics, hence a better understanding of the problem itself. As Desain et al. (1998) argue, a computational model should not be seen as "an aim unto itself" but rather as a "starting point of analysis" and "a means to compare and communicate theories between different research communities."

---

[1]For instance, Desain et al. (1998) argue that evaluating computational models by comparing their global output behaviors to that of humans "is too coarse to make substantial claims about the psychological validity of the model," and that we should "open up" the model and compare models and humans at the finer scale of subprocesses responsible for parts of the global behavior. For a critical discussion of this argument, see (Scheirer, 2000, pp.63-67).

In sum, we acknowledge that the relatively good performances of the computer programs proposed in this book are not a sufficient condition to make claims about human perception of rhythm. We hope, however, that the diverse elements of discussion raised while proposing implementation choices for these programs can provide insights for theories of rhythm perception.

1.3 Aims and book outline

General aims for this book are the following:

1. Submit a general framework for the qualitative comparison of rhythm description systems.

2. Provide an exhaustive review of existing computational attempts to rhythm description.

3. Highlight current promising research directions in computational rhythm description.

4. Determine which low-level audio features are best suited to the computation of useful rhythm periodicity functions.

5. Illustrate what rhythm periodicity functions are useful for via the design of algorithms for tempo induction, tatum estimation, time signature determination, swing estimation and musical genre classification.

The remainder of this book is organized as follows.

Chapter 2 covers the background. We present rhythmic concepts of first relevance to the understanding of this book, namely the metrical structure, tempo and timing. We propose a unifying framework for computational approaches to rhythm description and review existing systems (including systems for tempo induction, beat tracking, quantization and many others) with respect to the functional units of the proposed framework. In addition to this qualitative survey, we propose a quantitative comparison of state-of-the-art audio tempo induction algorithms. In the end of this chapter, we highlight recent achievements in automatic rhythm description and submit (on page 82) a list of open issues (with a special focus on tempo induction issues) which we believe make up the main promising current research directions.

Chapter 3 addresses one of the current research topics highlighted in Chapter 2: the determination of the low-level features of musical audio signal that convey best the predominant information relevant to rhythmic analyses. We make the assumption that the low-level audio features that are adequate for the computational identification of beat positions are *also* promising features for the computation of useful periodicity functions. The method used in Chapter 3 is based on this assumption: we analyze beat-labeled audio data and seek features whose temporal behavior would best indicate the presence and localization of beats.

In Chapter 4, we illustrate the usefulness of the features selected in Chapter 3 in the task of tempo induction. We also address several current open issues in the area of tempo induction, as the choice of periodicity function, the strategy for combining and parsing multiple information sources and whether the joint estimation of several metrical levels helps the determination of tempo.

In Chapter 5, we shortly introduce the research area of music content processing and music information retrieval and illustrate the use of rhythm periodicity functions and descriptors derived from such functions for music content processing: on the one hand in genre classification experiments and on the other hand in content-based transformations.

Finally, Chapter 6 summarizes the contributions of this book and proposes paths for future research.

Chapter 2

Survey of existing approaches to rhythm description

In this chapter, we first present rhythmic concepts of first relevance to the understanding of this book, namely the metrical structure, tempo and timing. In Section 2.2, we propose an unifying framework for computational approaches to rhythm description and review existing systems (including systems for tempo induction, beat tracking, quantization and many others) with respect to the functional units of the proposed framework. In addition to this qualitative survey, we propose in Section 2.3 a quantitative comparison of state-of-the-art audio tempo induction algorithms. Finally, the last section of this chapter concludes on recent achievements in automatic rhythm description and submits a list of open issues (with a special focus on tempo induction issues) which we believe make up the main promising current research directions.[1]

2.1 Representing musical rhythm

According to Fraisse (1982, p.149), "a precise, generally accepted definition of rhythm does not exist." The same consideration can be found in the very introduction of

---

[1]Part of the material in this chapter was previously published as stand-alone papers (Gouyon and Meudic, 2003; Gouyon and Dixon, 2005; Gouyon et al., 2006) and an academic report (Gouyon, 2003). Coauthors of the papers are thanked for their collaboration.

(London, 2005). In this book, "rhythm" is another term for "musical time," we will
not provide a more specific definition. Just as it has a melodic dimension, music has
a rhythmic dimension. Here, the word "rhythmic" implicitly encompasses small- and
large-scale temporal phenomena.

It is indeed difficult to draw a clear line regarding the temporal scope of rhythm.
Some use this word when referring to the duration of a note, expressed relatively to a
reference pulse, as in "this note is an eighth-note" (see Section 2.2.5). Others refer to
rhythm as a pattern of notes, as e.g. a typical Waltz pattern. For Cooper and Meyer
(1960), prosody defines the possible rhythmic patterns (*iamb, anapest, trochee, dactyl*
and *amphibrach*). Others use the related adjective — "rhythmic" — to describe somehow a percept that leads (or not) to dance to the music. Rhythm is commonly
defined indirectly. For instance, when stating that rhythm involves regularity (or
organization) and also differentiation (Fraisse, 1982, p.151), it is often opposed to the
"meter" (defined in more details below) and the "form." The three terms involve
regularity and differentiation, yet, the distinction lies in the concept of "perceptual
present," introduced by Fraisse (1982): "the temporal extent of stimulations that can
be perceived at a given time, without the intervention of rehearsal during or after
the stimulation." (Clarke, 1999, p.474). For London (2005), "rhythm involves the
pattern of durations that is phenomenally present in the music, while meter involves
our perception and anticipation of such patterns." He also puts it differently: "meter
[is] a mode of attending, while rhythm is that to which we attend." London considers that rhythm's proper meaning refers to the "smaller-scale features of musical
experience." The reason for this would be that rhythm "is apprehended within the
span of the perceptual present," unlike the form and the meter that would "engage
one's long-term memory of the piece at hand as well as one's musical background
and knowledge." Similarly, Clarke (1999) makes the distinction between "small- to
medium- scale temporal phenomena" (rhythm) and "large-scale temporal phenomena" (form). However, according to Cooper and Meyer (1960, p.6) rhythm extends
to all scales of temporal phenomena (from single note to entire movement). Confusion
abounds. But even if it seems futile to seek a more accurate definition of rhythm than
the broad "musical time," it is possible to define concepts that are related to it. This

is what we intend to do in the rest of this section.

Now, imagine the following musical scene. Somebody (or some machine) is making music: musical events are generated at given instants. A naive approach to describe the rhythm of this musical data (whether audio or symbolic) is to specify an exhaustive and accurate list of onset times, maybe together with some other musical features characterizing those events (e.g., durations, pitches and intensities in the MIDI representation). However, such a representation lacks abstraction. There is more to rhythm than the absolute timings of successive musical events. There seems to be agreement on the fact that, in addition, one must also take into account the *metrical structure, tempo* and *timing* (Honing, 2001). However, there is no consensus regarding explicit representations of these three rhythmic concepts.

A first reason for this lack of consensus is that different rhythmic features are relevant at each step in the *musical communication chain*, at each step where rhythmic content is produced, transmitted and/or received. As we illustrate in the next sections, metrical structure, tempo and timing take slightly different meanings for composers, performers and listeners. Indeed, even if a goal in the field of music psychology is to seek representational elements, or processes, that would stand as "universal" or "innate" (i.e. functioning at birth, independent of environmental influence) (Drake and Bertrand, 2001), a more widespread objective is to determine differences in perception according to listeners' culture, musical background, age or sex (Drake, 1993; Lapidaki, 1996, 2000; Gabrielsson, 1973a,b; Drake et al., 2000b).

A second reason is that the diverse *media* used for rhythm transmission suffer a trade-off between the level of abstraction and the comprehensiveness of the representation. Standard Western music notation provides an accepted method for communicating a composition to a performer, but it has little value in representing the interpretation of a work as played in a concert. On the other hand, a MIDI file might be able to represent important aspects of a performance, but it does not provide the same level of abstraction as the score. At the extreme end, an acoustic signal implicitly contains all rhythmic aspects but provides no abstraction whatsoever. In an application context, the choice of a suitable representation is based on the levels of detail (respectively abstraction) of the various aspects of music which are provided

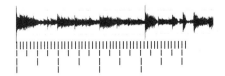

Figure 2.1: Representation of a four level metrical structure corresponding to an audio file.

by the representation (Gouyon and Meudic, 2003).

2.1.1 Metrical structure

Western music notation provides an objective regular temporal structure underlying musical event occurrences and organizing them into a hierarchical metrical structure. This is independent of the hierarchical phrase structure which may be explicit in the notation or implicit in the composer's, the performer's and/or the listener's conceptualization of the music.

The Generative Theory of Tonal Music (GTTM) formalizes this distinction by defining rules for a "musical grammar" which deals separately with grouping structure (phrasing) and metrical structure (Lerdahl and Jackendoff, 1983). While the grouping structure deals with time spans (durations), the metrical structure deals with durationless points in time, the *beats*.

Pulse — Beat   Cooper and Meyer (1960) define a pulse as "one of a series of regularly recurring, precisely equivalent stimuli. [...] Pulses mark off equal units in the temporal continuum." Commonly, "pulse" and "beat" are often used indistinctly and refer *both* to one element in such a series and to the whole series itself. Unlike Cooper and Meyer (1960), in this book, we use the term "pulse" to refer to a metrical level (that is, the whole series of beats at any metrical level) while the term "beat" refers to a single element of a pulse.[2]

Beats can be grouped together according to their respective accentuation (Cooper and Meyer,

---

[2]Further, "beat positions" or "beat indexes" will have the same meaning as "beats."

1960; Lerdahl and Jackendoff, 1983). "Accentuation" commonly refers to the human ability to perceptually apply a mark on some events in the musical flow, in opposition to other such events. The way we actually carry out this marking process is still not well understood. Pitch, intensity, duration (Dixon and Cambouropoulos, 2000; Snyder and Krumhansl, 2001), harmony (Lerdahl and Jackendoff, 1983) and timbre perception certainly have an influence on our way to hear rhythm in music (Thiemel, 2005). But one cannot state unequivocally that one of these factors prevails, nor that these are the sole factors of rhythm perception. London (2005) defines an accent as a "means of differentiating events and thus giving them a sense of shape or organization."[3]

According to the GTTM (Lerdahl and Jackendoff, 1983), beats obey the following rules. Beats must be equally spaced. A division according to a specific duration corresponds to a *metrical level*. Several levels coexist, from low levels (small time divisions) to high levels (longer time divisions). There must be a beat of the metrical structure for every note in a musical sequence. A beat at a high level must also be a beat at each lower level. At any metrical level, a beat which is also a beat at the next higher level is called a downbeat, and other beats are called upbeats. Beats obey a discrete time grid, with time intervals all being multiples of a common duration, the smallest metrical level. See Figure 2.1.

Pulse period and phase A pulse is characterized by a period and a phase. Its period is the distance between two consecutive beats (also called the inter-beat interval, IBI, it is inversely proportional to the tempo of the metrical level it defines, see on page 23) and its phase is specified by the temporal location of one beat (usually the first beat).

Non-beat For the purpose of this book, especially Section 3, we define the term "non-beat" as any point in time that is not a beat at any metrical level.[4]

---

[3]See http://www.music.indiana.edu/som/courses/rhythm/illustrations/accent.html for related definitions of accentuation.

[4]Note that non-beats do not necessarily coincide with notes.

Tatum   The metrical structure smallest level lacks a commonly accepted name. Schloss (1985) refers to the "attack-point." Bilmes (1993) uses the term "tatum." In our understanding, Parncutt's "basic time unit" (1994) and Hofmann-Engl's "chronota" (2002) refer to the same concept. In (Gouyon et al., 2002) we used the term "tick." In this book, we will use the term "tatum." The tatum is better defined as "the regular time division that most highly coincides with all note onsets" (Bilmes, 1993, p.22) than as the shortest interval between notes. Indeed, in syncopated musical excerpt for example, the tatum may not be explicit in the list of successive note intervals, it may rather be implied by the relationships between those intervals, see (Gouyon et al., 2002, p.397) and Figure 5.1.

Time signature — Measure   Restricting the notion of meter to two levels, Yeston (1976) defines it as "an outgrowth of the interaction of two distinct levels (two differently-rated strata), the faster of which provides the elements and the slower of which groups them." This definition seems close to the usual description found in a score, given by the time signature and the bar lines. The bar lines define the slower of the two levels (the "measure") and the time signature defines the number of faster beats that make up one measure. For instance, a $\frac{6}{8}$ time signature indicates that the basic temporal unit is an eighth-note (a "note" referring to a "whole," or "semi-breve") and that between two bar lines there is room for six of them. Two categories of meter are generally distinguished: duple and triple. This notion is contained in the numerator of the time signature: if the numerator is a multiple of two, then the meter is duple, if not a multiple of two but of three, the meter is triple. For instance, $\frac{2}{4}$ and $\frac{4}{4}$ signatures are duple, $\frac{3}{4}$ and $\frac{9}{8}$ are triple.

Quantized duration   The GTTM specifies that there must be a beat of the metrical structure for every note. Accordingly, given a list of note onsets, the quantization (or "rhythm parsing," see Section 2.2.5) task aims at making it fit into Western music notation. Viable time points (metrical points) are those defined by the different co-existing pulses. Quantized durations are then rational numbers (e.g. 1, $\frac{1}{4}$, $\frac{1}{6}$) relative to a chosen time interval: the time signature denominator. However, quantized durations can always be the object of controversy. For this reason, Cemgil et al. (2000) define quantization as "the extraction of an *acceptable* description (music notation) from a music performance" (original emphasis), "acceptable" meaning easy to read while representing the timing information accurately.

Dynamic Attending Theory Music psychology research asserts that humans perceive part of the metrical structure. Drake and Bertrand (2001) advocate a universal "predisposition toward simple duration ratio," and claim that "we tend to hear a time interval as twice as long or short as previous intervals." The Dynamic Attending Theory (Jones and Boltz, 1989; Drake et al., 2000a) proposes that humans spontaneously focus on a "referent level" of periodicity, and they can later switch to other levels to track events occurring at different time spans (for instance, longer-span harmony changes, or a particular shorter-span fast motive). However, metrical structure perception is strongly dependent on musical training (Drake et al., 2000b).

2.1.2 Tempo — Tactus

The tempo refers to the pace of a musical excerpt (how fast or slow it is). Given a metrical structure, *tempo* is defined as the rate of the beats at a given metrical level, for example the quarter note level in the score. It is inversely proportional to the pulse period. Here, the pulse can correspond to the score temporal unit, in this case, one refers to M.M. tempo (Maelzel metronome); Drake et al. (1999) refer to the "musical tempo." There is usually a *preferred* or *primary metrical level*, which corresponds to the rate at which most people would tap or clap in time with the music, and this is commonly used to define the tempo, expressed either as a number of beats per minute (BPM), or as the inter-beat interval. In many cases the primary metrical level corresponds to the denominator of the time signature, and the next one or two higher levels are specified by the numerator of the time signature. In this book, we use the term "tactus" to refer to the tempo of the perceptually most prominent pulse, (Lerdahl and Jackendoff, 1983).

However, it is not always correct to assume that the denominator of the time signature corresponds to the "foot-tapping" rate, nor to the "physical tempo" that would be an inherent property of musical flows (Drake et al., 1999). Human anatomy and motor-behavior naturally account for pulses (walking, heartbeat, breathing, brain waves, etc.). It is commonly thought that there is an intimate connection between these physiological properties and human perception of rhythm (Fraisse, 1982, pp.151-155), (Lapidaki, 1996, pp.40-47). We tend to perceive them as regular (Large and Palmer, 2002). Even when the sequences are not regular (Madison and Merker, 2002), or do not have rhythmic "intentions" (Iyer, 1998), as e.g. ocean waves (see the concept of "subjective rhythmisation" (Fraisse, 1982, p.155)). Drake and Bertrand (2001) argue that our "predisposition towards regularity" should be regarded as an universal of music temporal processing. They argue that "processing is better for regular than irregular sequences. We tend to hear as regular sequences that are not really regular." We would also "spontaneously search for temporal regularity and organize events around the perceived regularity." Some researchers propose that this regularity perception would be consistent over time (Clynes and Walker, 1986), (Clynes and Walker, 1982, p.188) and also independent of musical training (Levitin and Cook, 1996), arguing that our memory would store *absolute* tempo. Others rather consider that this would be true for few special cases (as when the sequences are well-known, or for a relatively restricted number of persons, as e.g. professional musicians), the general case being that regularity perception would be an unstable feature, relative to many factors: age, musical training, musical preferences, general listening context (as e.g. tempo of a previously heard sequence, subject's activity, instant of the day), etc. (Lapidaki, 1996, 2000; Drake et al., 2000a; Drake, 1993; Drake et al., 2000a; McAuley and Semple, 1999).

Nevertheless, differences in tempo perception are far from random; they most often correspond to a focus on a *different metrical level*, e.g. differences of half or twice the inter-beat interval (when hearing duple meter music) or one-third or three times the inter-beat interval (when hearing triple or compound meter music).

Preferred tempo and perceived tempo Regularity perception necessarily has upper and lower boundaries (e.g. Moelants (2002) proposes 1500 and 200 ms, respectively). They are imposed by the mechanical capacities of our perceptual apparatus and our short-term memory limits. Parncutt (1994, p.424) refers to an "existence region of pulse sensation." In between the boundaries, durations are all possible candidates to a perceived regularity, but *with different probabilities*. That is, we consider tempi with some "a priori" preference, this, independently of the auditory stimulus. Tempo preference distributions are commonly modeled as unimodal distributions (e.g. a Gaussian) with a maximum, for Fraisse (1982) and Parncutt (1994, p.438), at 600 ms. Moelants (2002) proposes a resonance curve (as that of a physical resonator) with a resonance period of 480 ms. Some also propose to consider multimodal distributions (Drake et al., 1999, p.201). Drake and Bertrand (2001) argue that a "temporal zone of optimal processing" around 600 ms may be considered as another universal feature.

Furthermore, when hearing a musical excerpt, one can give an appreciation whether it is fast or slow without referring explicitly to a specific pulse (Drake et al. (1999) refer to the "perceived tempo"); in addition to the perception of a specific pulse, the rapidity of an excerpt relies on the perception of "event density."

2.1.3 Timing

Although it is supposed to model the listener's intuitions, a major weakness of the GTTM is that it does not deal with the departures from strict metrical timing which are apparent in almost all styles of music. Thus it is only really suitable for representing the timing structures of musical scores, or as an abstract representation of a performance, where the expressive timing is not represented.

There are conceptually two types of non-metrical timing, which come under the headings *tempo* and *timing* respectively. These are illustrated in Figure 2.2, which shows a strictly metrical (isochronous) pulse (A), followed by three variations on this pulse. There are two types of timing changes: in the first case (B), just one beat in the pulse is displaced, whereas in the second case (C), all beats from a particular time onwards are displaced, as when a pause occurs in the music. In both of these cases,

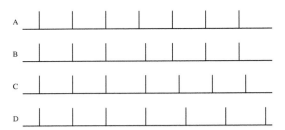

Figure 2.2: Four time-lines, marked with onsets, illustrating the difference between tempo and timing changes: (A) an isochronous pulse; (B) a local timing change; (C) a global timing change; and (D) a tempo change.

the change is in the timing; there is a discontinuity in the pulse, but the rate of the pulse on both sides of the discontinuity is the same. In this sense we can associate timing changes with short term changes in the pulse. On the other hand, a tempo change is a change in the rate of the pulse (D), which is a long term change in the pulse.

It is important to note that at the time of the first change (the 4th beat), it is impossible to distinguish cases (B), (C) and (D). This makes causal analysis impossible (i.e. algorithms which do not use information about future events in analyzing present events, as, for example, any real-time algorithm), since with no knowledge of the future, a single "out of time" beat could be due to either a tempo or timing change (Cambouropoulos et al., 2001).

One of the greatest difficulties in analyzing performance data is that the two dimensions of tempo and timing are projected onto the single dimension of time. Mathematically, it is possible to represent any tempo change as a series of timing changes and any timing change as a series of tempo changes, but these descriptions are somewhat counterintuitive (Honing, 2001). The parsimony of the representation is an important factor in its psychological plausibility (Tanguiane, 1993).

Tempo curves In order to represent changing tempi, various approaches can be used. If tempo is considered as an instantaneous value, it can be calculated as the inter-beat interval measured between each pair of successive beats. A more perceptually plausible approach is to take an average tempo measured over a longer period of time. A measure of central tendency of tempo over a complete musical excerpt is called the *basic tempo* (Repp, 1994), which is the implied tempo around which the expressive tempo varies. The end result of any of these approaches is a value of tempo as a function of time, which is called a tempo curve. Often, timing is also modeled by the tempo curve representation, an approach which is sharply criticized by Desain and Honing (1991) and Honing (2001) for failing to separate the dimensions of tempo and timing. This criticism is well supported by examples where transformations applied to a tempo curve representation do not preserve musically important features (Honing, 2005).

Recent research indicates that perceived beats do not necessarily line up exactly with onsets of musical tones, our perception rather favoring smooth tempo curves (Dixon, 2005).

Systematic deviations Among others, Bilmes (1993) and Baggi (1991) propose to represent timing deviations as systematic event shifts occurring within the span of the fastest pulse, while keeping a constant execution speed. They found evidence of the suitability of such a representation in analyzing respectively Latin percussion music and Jazz music. Friberg and Sundström (1999, 2002) propose to focus on the swing.

The term "swing" originates in jazz music. For Friberg and Sundström (2002), one characteristic aspect of the swing is that "consecutive eighth-notes are performed as long-short patterns." Laroche (2001) defines it as a "slight delay of the second and fourth quarter-beats" (in his article, "beats" refers to beats at the half-note level). The swing ratio refers to the ratio of the first eighth-note duration divided by that of the second.

The term "groove" resists precise definitions, but, as the "feel," it usually refers to a rhythmic phenomenon, resulting from the conflict between a fixed pulse and various timing accents played against it; or resulting from the "musician moving in non-metronomical ways" (Waadeland, 2001). The swing (as defined above) is a particular case of groove.

Categorical perception of deviations   Music psychology research presents evidence that listeners perceive performers' intentional timing deviations. Clarke (1987) shows that "categorical perception" differentiates expressive timing from rhythmic structure: a small number of categories are used to characterize the *continuously* variable temporal transformation of the *discrete* (integer ratio) structure. Further, timing and structure are tightly linked. Repp (1992) confirms listeners' sensitivity to timing deviations, but, most importantly, also shows that this sensitivity is a variable of the position in the metrical structure. Complementary to this finding, there is strong evidence that performers do not produce timing deviations at arbitrary points in time (Palmer, 1997). They rather deviate from pure mechanical performance in specific ways. The metrical structure provides "anchor points" for timing deviations, and "every aspect of musical structure contributes to the specification of an expressive profile for a piece" (Clarke, 1999, p.492). Expressive timing is also *systematic*; the timing in repeated performances can be very stable over a period of years (Clynes and Walker, 1982, pp.181-187).

2.2   Computational rhythm description

The chief goal in automatic rhythm description is the parsing of acoustic events that occur in time into the more abstract notions of metrical structure, tempo and timing, as illustrated in Figure 2.3, where the goal is to derive a representation like (B) from (A) or (A'). A major difficulty is the inherent ambiguity of rhythm, as discussed in the previous section. This concern is also expressed by Parncutt (1994, p.423), Chung (1989, p.19) and Rosenthal (1992, p.12). This is a problem because computer implementations demand precise definitions, and any systematic comparison of program performances must be based on some "ground truth." For instance, it is difficult to

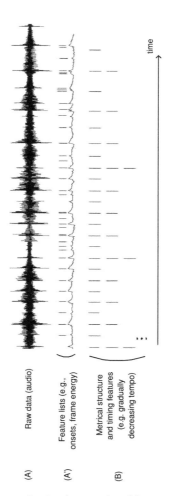

Figure 2.3: Example of an audio signal, examples of feature lists (onsets, short-term energy), corresponding metrical structure and timing features, showing a gradually decreasing tempo.

compare programs that extract the tempo if their definitions of tempo do not explicitly refer to the same metrical level. Indeed, tempo induction systems typically make
errors of simple integer ratios, such as 60 BPM instead of 120 (Goto and Muraoka,
1997; Dixon, 2001a; Gouyon et al., 2006). Also, even if existing scores can be taken
as ground-truth references to the quantization or time signature determination tasks,
"correct" time signatures or quantized durations can always be the object of controversy (see on page 22).

The ambiguity of rhythm representations becomes apparent when we consider the
following questions: Given a musical signal, how many metrical levels are relevant?
Is there one most important level? Is there solely one *correct* tactus? Which metrical
level defines the M.M. tempo of the music? Which metrical levels define the time
signature? What are the relevant categories of timing deviations? In terms of Figure 2.3, what elements of (B) are relevant, and how can they be named and clearly
defined? Are the answers to these questions common to all listeners?

These questions have no simple answers. There is no canonical form for representing rhythm, and lacking this ground truth, it is difficult, if not impossible, to
provide a meaningful quantitative comparison of the various computer systems which
each have different answers to these questions. Further, until recently, there were no
common database on which systems could be tested (see Section 2.3 for more details
on systematic evaluations).

Some systems derive the beats and the tempo of just one metrical level, where
this level is somewhat arbitrarily chosen. Others aim at deriving complete rhythmic
transcriptions (i.e. score rhythm representations) from musical performances. Still
other programs aim at determining some timing features from musical performances,
such as tempo changes, event shifts (timing changes) and swing factors.

These computer programs share some functional aspects. For instance, a preva-
lent aspect is the handling of symbolic or parametric data derived from (or instead
of) raw audio data. These feature lists are usually made up of onset times (see (A') in
Figure 2.3), which are sometimes used in conjunction with other features (temporal,
timbral, harmonic or melodic). We define feature lists somewhat broadly, to include

frame-based feature vectors as well as lists of parameterized events, since the algorithms subsequently used to process the lists are similar for both cases, even though the time scales differ by an order of magnitude. The distinction between high-level and low-level representations, although conceptually important, does not necessarily play a large role in determining the suitability of algorithms for the discovery of temporal patterns.

This section provides a qualitative comparison of systems with respect to the functional units of the general model illustrated in Figure 2.4, consisting of feature list creation (e.g., onset detection), pulse induction (including periodicity computation, pulse selection, handling of event shifts and strategies for combining and parsing multiple information sources), pulse tracking, time signature, quantized duration and rhythm pattern determination, estimation of short term timing features and extraction of periodicity features via parameterization of periodicity functions. In the remainder of this section, we discuss each of these functional units in turn.

2.2.1 Feature list creation

Some computer systems deal with symbolic or parametric data, such as manually parsed scores or MIDI data containing solely onset times and durations (Brown, 1993; Longuet-Higgins and Lee, 1982). Recent systems tend to deal directly with acoustic signals or with compressed audio (Wang and Vilermo, 2001),although some early systems also used audio input (Chowning et al., 1984; Schloss, 1985). No matter what input data is used, the first analysis step is the creation of a feature list, i.e. the parsing, or "filtering," of the data at hand into a sequence of features. These features range from note onset features (as time, duration and amplitude) to frame-based signal features, and they are assumed to convey the predominant information relevant to rhythmic analysis.

In this step, monophonic excerpts are often parsed into sequences that resemble note features (e.g. onset time, duration and pitch). For polyphonic music, Allen and Dannenberg (1990) propose separating instrumental streams (a very challenging goal) and building a feature list for each monophonic stream, which is merged

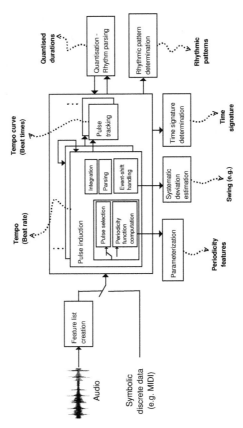

Figure 2.4: Functional units of rhythm description systems. Each functional block is the object of detailed explanations in the text.

with other streams after some rhythmic analysis steps. Another possibility is to describe a polyphonic excerpt by a single feature list, giving a global (homophonic) view where features ("summary events" in (Rosenthal, 1992, p.29)) represent musical chunks such as chords or energy components.

2.2.1.1 Event-wise features

Onset time   The extraction of note onset times for rhythmic analysis is ubiquitous in the literature. Musical event occurrence instants are very important cues for rhythm perception. Onsets can be extracted (with more or less reliability) from virtually any musical format. For instance, Longuet-Higgins (1987) and Brown (1993) process onsets manually parsed from scores. They can also be easily parsed from MIDI data (Cemgil et al., 2000; Raphael, 2002; Dixon and Cambouropoulos, 2000; Cemgil et al., 2001). More complex is their automatic extraction from audio signals, the details of which are out of the scope of this book. Early systems for musical onset detection include (Chowning et al., 1984), more recent systems include (Klapuri, 1999) and (Thornburg and Gouyon, 2000). Bello (2003) provides an exhaustive overview of musical onset detection.

Duration   In addition to onset times, some systems also handle durations, or alternatively inter-onset intervals (IOIs), which can be considered as roughly equivalent to durations (Brown, 1993), and are easier to compute for audio data. Like onsets, durations can be easily parsed from MIDI or scores, but cannot be computed reliably from audio data (especially polyphonic music). Durations are extracted from scores by Brown (1993) and Brown (1993); Longuet-Higgins and Lee (1982), whereas Mont-Reynaud and Goldstein (1985), Dannenberg and Mont-Reynaud (1987) and Allen and Danne (1990) use durations derived from MIDI data to filter out "weak" onsets, being those onsets whose duration is either shorter than some fixed threshold (20 to 50 ms for Allen and Dannenberg (1990) or much shorter than the preceding one. Chung's "note importance agencies" (1989, pp.61-62) and Temperley and Sleator model (1999) also parse MIDI data into note onsets and durations. Parncutt (1994, p.426-432) weights onsets proportionally to their subsequent IOI, using a perceptually justified "satura-tion function." Perceptual experiments by Snyder and Krumhansl (2001) show that timing information alone (onsets and durations) is sufficient for the perception of a pulse in Ragtime music.

Timing patterns   Mont-Reynaud and Goldstein (1985) focus on repetitions of note timing patterns. Dealing with scores, Mont-Reynaud and Goldstein (1985) demon-strate the difficulty of the pattern elaboration process and its inherent multiplicity of solutions. Lartillot's technique (2004) for discovering recurrent patterns in symbolic music sequences could be used here. However, it is difficult to envisage this rationale when using audio signals (all the more if polyphonic).

Relative Amplitude   Relative amplitude is a factor which contributes to percep-tual accentuation, and is easily computable from MIDI or audio data, but it is largely absent from score notation. Various systems use relative amplitude for weighting on-sets derived from symbolic (Smith and Kovesi, 1996; Smith, 1996; Dixon and Cambouropoulos, 2000; Gasser et al., 1999) or audio data (Dixon, 2001a; Dixon et al., 2003; Gouyon et al., 2002).

Pitch   Pitch is easily obtained from scores and MIDI data, but still cannot be reli-ably extracted from polyphonic audio (Gómez et al., 2003; Klapuri, 2004). Pitch is rarely used in automatic rhythmic analysis. Chowning et al. (1984) extract from au-dio signals of simple melodic lines played on a piano note onsets, durations and pitches. This data makes up the "acoustic maps" that are subsequently processed in order to determine complete score transcriptions of musical performances. Dixon and Cambouropoulos (2000) extract duration, amplitude and pitch from MIDI data in order to calculate the "salience" of musical events, which is shown to improve the performance of their beat tracking system (when events are made up of several notes, the longest duration, the amplitude summation and the lowest pitch are kept as representative).

Chords   Chords are used in two ways in rhythmic analysis: by counting the number of simultaneous notes as a measure of accentuation (Dixon, 2001a; Rosenthal, 1992), and by detecting harmonic change as evidence of a downbeat (Goto and Muraoka, 1999) and (Temperley and Sleator, 1999, p.25). Just like pitch, chords are easily readable in scores and MIDI data, but much harder to derive from audio data (Gómez, 2005).

Percussive instrument classes Percussive events can be extracted from MIDI (MIDI channel 10 is normally used for such events). Their extraction from audio data is still ongoing research.

Schloss (1985) differentiates between "high drums" and "low drums." Similarly, Bilmes (1993) proposes to automatically differentiate several conga sounds. As of today, isolated percussion samples can be automatically classified with a high reliability (Herrera et al., 2003).

In the more complex task of transcribing audio drum tracks,[5] recent research have seen some progress (Gouyon and Herrera, 2001; FitzGerald et al., 2002; Paulus and Klapuri, 2003; Gillet and Richard; 2004).

However, the state-of-the-art in percussive events recognition in polyphonic mixtures leaves room for improvement. Goto and Muraoka (1995) deals with audio signals whose tactus is maintained by specific drum sounds and in addition to onset detection, a discrimination is made between bass drums and snare drums. The classification is based on template-matching of the spectrum of short regions surrounding onsets. Instead of addressing drum classification with models (or templates) built in a training phase, recent research tends to adapt such models to the characteristics of each audio signal at hand in order to match closely, in an iterative process, the actual recurrent percussive timbre of the audio signal (Gouyon, 2000; Gouyon et al., 2000; Zils et al., 2002; Yoshii et al., 2004; Sandvold et al., 2004).

2.2.1.2 Frame-wise features

Honing (1993) comments that "there seems to be a general consensus on the notion of discrete elements (e.g. notes, sound events or objects) as the primitives of music ... but a detailed discussion and argument for this assumption is missing from the

---

[5]a drum track is an audio signal containing solely drum samples

literature." Further, Scheirer (2000) argues that solely well-trained musicians hear the music in terms of its conventional musicological structures, and he criticizes the "transcriptive metaphor," maintaining that the modeling of the perceptual mechanism should not be based upon abstract symbols such as durations, pitches, and chords. For example, he showed in an informal experiment that replacing the harmonic content of a musical signal with modulated noise did not change the sensation of tempo (Scheirer, 1998).

Based on this rationale, some systems do not focus on note onsets and their features, but refer to a data granularity of a lower level of abstraction: frames. A frame is a short chunk (typically 20 ms) of audio, from which both time and frequency domain features can be computed. Consecutive frames are usually considered with some overlap for smoother analyses. The analysis step, the *hop size* (typically 10 ms), equals the frame size minus the overlap.

Energy   The simplest feature is energy, which can be calculated for the whole frame or for frequency subbands of the frame. Assuming that low-frequency instruments communicate much of the rhythmic information, Alghoniemy and Tewfik (1999) and Blum et al. (1999) focus on the energy in low-frequency components, as a simple alternative to the percussion detection methods mentioned previously. Others decompose the signal into several subbands, compute energy in each subband, postprocess them (e.g. assign them different weights) and then sum them back (Vercoe, 1997; Tzanetakis and Cook, 2002). Finally, another procedure is to compute one feature list per frequency subband, yielding e.g. 6 feature lists for Herre et al. (2002) and Wang and Vilermo (2001) (in the latter, MP3 bitstreams are processed, hence frames are 13 ms-long and there is no overlap, the correspondence between subband frequency interval and MDCT coefficients depends on whether short or long windows have been used in the MP3 coding), 20 for Pampalk et al. (2002) and 23 for Sethares and Staley (2001).

Energy variation   Rather than focusing on frame energy values, some systems measure the variation of the energy between consecutive frames. For instance, Foote and Uchihashi

(2001) use the cosine distance between the magnitude spectra of consecutive frames (11 ms-long, no overlap). In (Laroche, 2003), frame magnitude spectrum (10 ms frames, no overlap) is transformed by a compression function (e.g. a hyperbolic sinus) to give higher weights to high frequencies than low frequencies, and then a first-order difference is computed. Scheirer (1998) also computes the first-order difference of frame energy values in 6 frequency bands.[6] In (Klapuri et al., 2005), the computation of "registral accents" entails the aggregation of the energy values computed in 36 frequency bands in a smaller set of feature lists (e.g. 4). Here also, a first-order difference replaces the frame energy value.

One might note that these procedures resemble the first stages of an onset detector. The main difference is that there is no discretization of frame energy values, nor any explicit thresholding and peak-picking. Further rhythm description stages deal with a data granularity defined by the hop size.

Other low-level features   Low-level features other than energy (e.g. spectral flatness, temporal centroid) have also been recently advocated (Gouyon and Herrera, 2003b).

2.2.1.3   IBI-wise features

Several authors propose to compute low-level features over the time span defined by two consecutive beats at a given metrical level. For instance, Seppänen (2001) and Gouyon and Herrera (2003a) compute beat indexes from low level features computed on segments of audio defined by the smallest metrical level, the tatum. Also, Goto and Muraoka (1999), Meudic (2002) and Gouyon and Herrera (2003b) derive downbeat indexes from descriptors of inter-beat segments. The latter points out the relevance of a specific feature for downbeat computation: the temporal centroid of inter-beat segments.

---

[6]It may be noted that Scheirer (1998) does not explicitly refer to a frame-by-frame analysis. However, in our understanding, the signal envelope extraction (by convolution with a half-Hanning window) and downsampling is similar to a framing of the signal (with a frame-size equal the window size, i.e. 200 ms) and a step of analysis (hop size) of 5 ms (if the downsampling frequency is 200 Hz; 13.3 ms when downsampling at 75 Hz, value advocated by Scheirer for reaching real-time performances, see on page 112).

2.2.2   Pulse induction and rhythm periodicity function computation

A pulse is defined as the periodic recurrence of a feature in time. Therefore, computer programs generally seek periodic behaviors in feature lists in order to select pulse periods and possibly also their phases (hence beats). The process of *pulse induction* aims at highlighting intrinsic periodicities of feature lists, and thus it is central to any form of rhythm understanding (see Figure 2.4).

The resulting beats often serve as input to a *pulse tracker* (see page 51). This division in the processing is motivated in Desain's "(de)composable theory of rhythm perception" (1992) that highlights the need to consider events with respect to the rhythmic context. This context can be defined mathematically as an expectancy curve, a function of past IOIs. Further, Desain and Honing (1999) argue that human perception of pulse exhibits two dichotomic processes: a bottom-up process that forms a pulse percept very rapidly from scratch, and a top-down process (a persistent mental framework) that lets this induced percept guide the organization of incoming events.

In pulse induction, a fundamental assumption is made: the pulse period (and phase) is *stable* over the data used for its computation. That is, there is no significant speed variation during the excerpt used for inducing a pulse. In that part of the data, remaining timing deviations (if any) are assumed to be short term (considered as either errors or expressiveness features). They are either "smoothed out" (see page 45) or cautiously handled within the pulse induction process so as to derive patterns of short term timing deviations as e.g. the swing (see page 58).

For pulse induction, computer programs either proceed by

- *pulse selection*, evaluating the importance, or salience (Parncutt, 1994), of a *restricted number* of possible periodicities (see page 40), or by

- *periodicity function computation*, generating a *continuous* function plotting pulse salience versus pulse period (or frequency) (see page 41).

The former procedure is simpler, and is typically used for processing symbolic data, where pulse selection is usually considered jointly with subsequent tracking. Systems handling finer-grained data (e.g. frame features) often implement a periodicity function computation.

Inducing the pulse with part of the data In many cases, a hypothesis is made on the maximum duration over which the pulse period can be considered stable (e.g. 5 s). In this case, the induction process serves as a front-end to a tracking process. The resulting pulse (there also might be several candidates) is propagated over the remaining data (i.e. for $t > 5\ s$) and a process of comparison between predicted beats and actual musical events produces all the beat positions and a tempo curve (see page 51). Most systems resorting to the pulse selection method (page 40) process a small amount of data for pulse induction and rather translate the overall difficulty onto the subsequent tracking process; e.g. reporting on potential problems of their induction technique, Allen and Dannenberg (1990) argue that it does not seem to be a problem since their tracking model "incorporates a great deal of flexibility." Some systems relying on the computation of a periodicity function also consider it as a first processing stage, previous to the pulse tracking (Dixon, 2001a; Rosenthal, 1992). Those typically use around 5 s of data for pulse induction; additionally, in some cases, some emphasis can also be given to most recent samples (e.g. by multiplying the data with an exponentially decreasing window, or by the intrinsic exponential behavior of a comb filter impulse response (Scheirer, 1998); the "tempogram" of Cemgil et al. (2001) also implements this feature in its parameter α).

This rationale is suitable for streaming application where one does not know a priori the amount of data to process.

Inducing the pulse with the whole data If the pulse induction process is achieved on the whole data (e.g. an entire audio recording, or MIDI file), a strong assumption is made, namely that the tempo is constant all over. Pulse tracking is simply not addressed in this case. This is suitable to some musical excerpts, but much music violate this assumption. This is typically done by systems that rely on a periodicity function computation.

This rationale is not suitable for streaming applications, but it may be relevant for specific off-line applications, where one knows that the the tempo stability assumption makes sense.

2.2.2.1 Pulse selection

The first approach to pulse selection is an instance-based approach, where each IOI defines a possible pulse period, and the corresponding events define the phase. For example, Longuet-Higgins and Lee (1982) simply consider the first two events as the first two beats, whereas Dannenberg and Mont-Reynaud (1982) take the first two agreeing IOIs as defining the pulse (they refer to this process as "creating a predictor"). In the system by Allen and Dannenberg (1990), the metrical value of the first event must be given, and the pulse is derived from this value and the first IOI. Chung (1989) derives a number of pulse periods and phases from the event list in a sequential manner. Like Longuet-Higgins and Lee (1982), Chung considers the first two events as potential beats. Subsequent events are considered in the light of this potential pulse: if they do not coincide with the pulse (after allowing some tolerance), a new potential pulse is created, its period being set to the most recent IOI, and the phase being specified by the current event. Limiting the number of pulses is achieved by assigning to each pulse a score depending on: the "importances" (i.e. durations) of its constituent events, the timing deviations of beats from expected beat positions and the number of syncopations. Solely the two or three highest-scoring pulses are selected. Chung (1989, p.77) reports that the system usually finds all relevant pulses within the first few bars. In sum, Chung's selection of pulse resembles that of Dannenberg and Mont-Reynaud (1987), improvements being that it is not restricted to two agreeing IOIs, and that more than one pulse are considered.[7]

It is also possible to seek periodic behaviors in the feature list by computing a

---

[7]This *sequential* pulse induction mechanism could be thought of as some kind of tracking (see page 51). But, in Chung's words, there is no beat tracking (1989, pp.61 and 87), the tempo is considered constant. Chung argues that if there were some tempo or time signature changes, his model would eventually discover it, as the pulse induction process is always running (previous "winning pulse" scores would diminish), but "there is no mechanism for expecting repeated changes" (p.88). This would finally result in a some *switching* of "agents" (and "agencies") rather than in a proper *evolution* of a single agent.

similarity measure between the list and several pulses. This procedure is foreshadowed by the "clock model" of Povel and Essens (1985), where "people perceive, remember and reproduce temporal patterns by structuring their representation according to an internal clock" (McAuley and Semple, 1999, p.178) with a period corresponding to the smallest IOI. This rationale is only suitable for parsed scores and artificially created sequences where IOIs are exact integer multiples of the clock period. In this case, goodness of fit between a pulse and an event list can be estimated by positive evidence (the number of events that coincide with beats), or by negative evidence (the number of beats with no corresponding event), or by combining these two counts (McAuley and Semple, 1999). Similarly, Parncutt (1994, pp.433-436) considers pulse induction on cyclically repeating impulse patterns, by expressing pulse period and phase with respect to the shortest IOI, and determining the "pulse-match salience" based on positive evidence rather than negative evidence.

2.2.2.2 Computing a periodicity function

The alternative to pulse selection is the computation of a periodicity function: a magnitude (or salience) corresponds to each period (or frequency) in the periodicity continuum. In practice, the range of periods is not continuous but sampled, with typical intervals being 5 or 10 ms. Some systems process several feature lists separately, for example by calculating periodicities in each frequency subband and then integrating the results (Scheirer, 1998; Paulus and Klapuri, 2002; Dixon et al., 2003; Gouyon and Herrera, 2003a).

The periodicity function may also be multiplied by a tempo preference (probability) distribution, e.g. (Parncutt, 1994, p.439, equation 7), implementing the fact that humans consider tempi with some a priori preference (see page 24).

Some methods also let slow periodicities affect faster, rationally-related, periodicities (e.g. a τ-periodicity in the feature list contributing to the raising of several peak magnitudes: naturally at τ, but also $\tau/2$, etc.), thus encoding aspects of the metrical hierarchy.

In some cases, an emphasis is given to the most recent samples, e.g. by multiplying the data with a decreasing window (Desain and de Vos, 1990),(Goto, 2001, equation

7), or by the intrinsic exponential behavior of a comb filter impulse response (Scheirer, 1998). The "tempogram" of Cemgil et al. (2001) also implements this feature in its parameter α.

Fourier transform   Periodicity functions are often calculated with standard signal processing algorithms, such as the Fourier transform, which Blum et al. (1999) applies to onset lists and Pampalk et al. (2002) uses on 20 frequency subbands of the audio signal.

Wavelets   Smith (1996) and Smith and Kovesi (1996) argue that wavelet analysis is well-adapted to capture temporal organizations at different scales and visualize the hierarchies between the different organizational levels. The choice of wavelet representation is not made to suggest that human perception actually proceeds by means of such signal representation; rather, "the intention is to make explicit that information which is inherent in the rhythm."

Auto-Correlation Function (ACF)   The most common signal processing technique for periodicity computation is the autocorrelation function (ACF), which has been applied to subband signals and to onset lists represented as Dirac delta functions (for scores or mechanical performances) or smoothed using e.g. a Gaussian function (to cater for small changes in timing and tempo; see Paragraph 2.2.2.3).

Brown (1993) computes a sample-by-sample ACF of a sequence of onsets sampled at 200 Hz, weighted by their durations. Her results are best for longer values of the integration time (the time span for the estimation of one correlation coefficient). The integration time is also important because it determines the statistical reliability of the estimate (Desain and de Vos, 1990). Scheirer and Slaney (1997) also compute the ACF of onset trains, and Scheirer (1997) advocates summing ACFs computed over several frequency channels.

The "Narrowed ACF" (NACF) was introduced by Brown and Puckette (1989): the coefficient at lag k is computed as the weighted sum of the ACF coefficients at lags which are integer multiples of k, where the weights decrease for larger multiples of k. The NACF implicitly encodes aspects of the metrical hierarchy (a $2k$-periodicity has an effect on the correlation coefficient of lag k) and gives better period precision at the expense of worse time resolution. Improved precision is a useful feature for signals that contain close periodicities, but this is an unlikely situation in the context of pulse induction. It may be noted that Brown (1993, p.1955) recognizes that the NACF is not necessary. Vercoe (1997) proposes the use of the "Phase-Preserving Narrowed Autocorrelation" in order to keep time localization normally lost in computing an ACF. The computation involves a simplified NACF, i.e. with a very short integration time, which reduces the stability of the estimate.[8]

Foote and Uchihashi (2001) propose two ways to compute periodicities ("self similarity") in feature lists: they build a similarity matrix and perform either sums or correlations of the matrix diagonal elements. The first of these two options resembles the computation of an ACF: the sum over the ith diagonal is similar to the (normalized) autocorrelation of the signal frame parameters with a lag i. The latter option is similar to the NACF, in that it goes further and accounts for aspects of the metrical hierarchy.

ACFs are also implemented by Gouyon et al. (2000), Goto (2001), Herre et al. (2002), Tzanetakis et al. (2002),Gouyon and Herrera (2003a), Gouyon and Herrera (2003b) and Dixon et al. (2003).

Comb filterbank An alternative approach uses a bank of resonators, each tuned to a possible periodicity, where the output of the resonator indicates the strength of that particular periodicity. This technique is foreshadowed by the "clock model" of Povel and Essens (1985) in that it seeks the series of periodically-spaced clock beats that best matches the feature list (the implementation in the form of a bank of comb filters introduces an exponential decay on the clock beat amplitudes). Scheirer (1998) uses comb filters as resonators, and performs periodicity analysis separately on 6 frequency subbands of the signal, and then sums the filterbank outputs across the subbands. 150 resonators are used to cover a logarithmically spaced frequency

---

[8]Interested readers should check the CSound implementation of the "Phase-Preserving Narrowed Autocorrelation" in the "tempest" method for tempo estimation (see http://www.lakewoodsound.com/csound/hypertext/manual.htm).

range from 1 Hz to 3 Hz (i.e. 60 to 180 BPM). Klapuri et al. (2005) also use a comb filterbank. Scheirer (1997, 1998) details similarities and differences between the NACF and comb filter approaches. This method also "encodes implicitly aspects of the rhythmic hierarchy" (Scheirer, 2000, p.91).

Time interval histogram   The use of histograms of time intervals between similar events is also widespread. These are typically IOI histograms, although Mont-Reynaud and Goldstein (1985) build histograms of time intervals between temporal *patterns* (see page 34), resembling somewhat an ACF. Chowning et al. (1984, pp.17-19) and Schloss (1985, p.90) generate a smoothed histogram by associating a Dirac delta function with each IOI, assigning it a weight proportional to its value (i.e. longer IOIs are emphasized) and convolving them with a "bell shaped curve of appropriate bandwidth." Similarly, Rosenthal (1992, p.40) builds a discrete IOI histogram and smears it with a Gaussian curve. The IOI clustering scheme of Dixon (2001a, 1999) is essentially similar to the building of an IOI histogram where the bins are not fixed. Clusters of similar IOIs are given scores based on the number of elements in the cluster and the amplitudes of their onsets. An adjustment of the scores (and cluster representative interval) then favors rationally-related clusters, thus encoding aspects of the metrical hierarchy. Seppänen (2001) and Gouyon et al. (2002) also implement IOI histograms. In the former, the computation is sequential and updated at each new event, emphasis being given to the most recent ones.

Periodicity transform   Sethares and Staley (2001) propose the "periodicity transform," which projects the signal (here made up of frame energy in a subband) onto a set of basis vectors. Unlike the Fourier and wavelet transforms, the basis vectors are not specified a priori, the transform rather finds the basis vectors which best match the signal.

Tempogram   Cemgil et al. (2001) define the "tempogram" which induces a probability distribution over the pairs {pulse period, pulse phase} given the onsets. Using a Bayesian framework, this probability (*posterior* distribution) is proportional to the *likelihood* of the observed onsets under given period and phase hypotheses, weighted by a *prior* distribution (which in this case is flat, as they consider all tempi to be initially equiprobable). For given periods and phases, the likelihood is computed as the integral, over all the onsets, of the product of a constant pulse (with appropriate period and phase) and a continuous representation of the onsets (onsets are smeared with a Gaussian curve). It implements the assumption that a good pulse is one which matches all the onsets well. The tempogram marginal probability function $p(w|t)$ (integral of the tempogram with respect to phase) provides a 1-dimensional representation of periodicities resembling those aforementioned (Cemgil et al., 2001, Figure 4).

Pulse matching Recall that the pulse selection method used by Parncutt (1994) and McAuley and Semple (1999) (see on page 40) is based the computation of a similarity measure between event lists and pulse templates. It can be generalized to deal with musical patterns which are not strictly metronomical and not cyclically repeating. In this case, the "basic time unit" is not known, but it is possible to enumerate all possible pulse periods and phases, as do Gouyon et al. (2002), who use both positive and negative evidence in the matching of onset lists and pulses (see on page 170).

If the feature representation is continuous (e.g. when adding some degree of tolerance for onset times by smearing, or when using frame energies), it is no longer meaningful to speak of positive and negative evidence. However, computing the inner product between pulses and the continuous feature list is possible (Laroche, 2003, equation 5). This resembles the aforementioned tempogram, the main differences are that the tempogram accounts for weights on beats and considers all phase candidates simultaneously.

2.2.2.3 Handling short-term deviations

Short-term timing deviations always exist in any musical data other than parsed scores and artificial sequences. Feature periodicities are always approximate. This is a problem especially when processing *discrete* event lists represented as a sum of

Dirac delta functions.

One solution is to consider events as having a "tolerance interval" (Longuet-Higgins, 1987). Dixon (2001a) uses a fixed tolerance interval of 25 ms (the "cluster width") for IOIs, whereas Dixon et al. (2003) and Chung (1989, p.65) employ tolerance intervals proportional to the IOIs, so that longer IOIs allow for greater variations. Seppänen (2001) quantizes the IOI histogram into a specific number of bins, giving a fixed tolerance interval, but does not state the number of bins. A tolerance interval can also be considered in the creation of the feature list, such as the "summary events" of Rosenthal (1992, p.29) which merge note events into chords if their onset times are within a timing tolerance of 10 ms. Similarly, Dixon and Cambouropoulos (2000) use a tolerance of 70 ms to define onset simultaneity.

The previous procedures can be interpreted as convolving the event list with a rectangular window. This helps in processing music with short-term timing deviations, but the resulting representation is still discontinuous (the sum of Dirac functions has been transformed into a step function). This can be improved by using smoother curves for smearing, such as a Gaussian window (Chowning et al., 1984; Schloss, 1985; Rosenthal, 1992; Cemgil et al., 2001; Gouyon et al., 2002), an exponential window (Dannenberg and Mont-Reynaud, 1987, p.245), or a triangular window (Tanguiane, 1994).

Some expressiveness timing features lie in short-term timing deviations. Therefore, instead of "smoothing them out," one may think of handling them cautiously to derive *patterns* such as the swing for instance (see page 58).

2.2.2.4   Combining multiple information sources

We highlighted in the previous sections different ways to compute periodicity functions and different low-level features of interest. In case several features are computed, the combination of these processing blocks into a pulse induction algorithm requires to make decisions regarding the following lines:

1. One may first integrate features and then compute a periodicity function on the resulting representation, or, inversely, compute a periodicity function for each

feature and then integrate these functions.

2. One may evaluate features and put different emphasis on them, or even select some features and discard some others.

3. One may also normalize features.

Regarding the first point, Scheirer (1998) claims that "a rhythmic processing algorithm should treat frequency bands separately, combining results at the end, rather than attempting to perform beat tracking on the sum of filterbank outputs."

Regarding the second point, it is possible to evaluate feature goodness, select some and discard some others before combining them; different criteria can be considered for the task, the most common being the periodic, or non-periodic behavior of features, which can be estimated from e.g. the ACF of a given feature by several approaches:

- The peakedness of the ACF. Intuitively, peaky ACFs should correspond to valuable features. This can be estimated via, for instance:

  - The variance of the ACF with respect to its linear fit.

  - The ratio of the maximum value and the mean value of the ACF (possibly after removing its linear fit)

- The periodicity of the ACF itself. Intuitively, periodic ACFs should correspond to valuable features.

In case one chooses to integrate information *before* periodicity function computation, the process, illustrated in Figure 2.5(a), becomes: feature normalization (optional) — feature evaluation and weighting/selection (optional) — combination of features — periodicity function computation.

In case one chooses to integrate information *after* periodicity function computation, the process, illustrated in Figure 2.5(b), becomes: feature normalization (optional) — periodicity function computation — periodicity function evaluation and weighting/selection (optional) — combination of periodicity functions.

The meaning of the word "combination" above may be the mathematical sum or product. Optionally, a weighted sum or product could be envisaged, however this

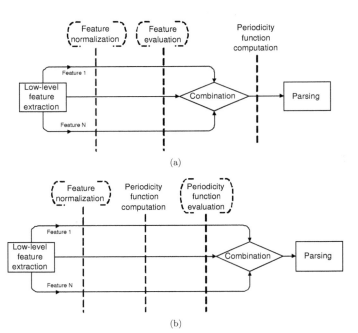

Figure 2.5: Different ways to combine multiple information sources: combining features before computing periodicity function (2.5(a)) or combining periodicity functions (2.5(b)).

would require an additional strategy for estimating the weights. Combining periodicity functions may also include broader concepts, and be considered together with periodicity function parsing (detailed in Paragraph 2.2.2.5). For instance, it is possible to select several prominent peaks of diverse periodicity functions and accumulate them in a histogram whose highest peak can be selected ((Tzanetakis and Cook, 2002), see also on page 64) or parse these peaks using some musical heuristics as detailed in the next paragraph.

2.2.2.5 Parsing the periodicity function

The desired output of the pulse induction process is a discrete pulse period (and optionally its phase) for each periodicity, rather than a continuous periodicity function. Therefore another step is needed in order to produce useful rhythmic information. Usually, this is achieved by a peak-picking algorithm such as an N-point running window method, which defines local maxima as points whose values are higher than those of their direct neighbors ($N/2$ on the left and $N/2$ on the right). Peaks must be subsequently interpreted with respect to their musical meaning, e.g. the tatum, tactus and measure periods, which can be identified using heuristics (Goto, 2001; Smith, 1996; Smith and Kovesi, 1996). As Smith and Kovesi (1996) put it, "while it is tempting to draw hypotheses for methods of derivation of the tactus by 'ridge-tracing' or the well-formedness of the global continuation of a voice, further research is required to build a model of tactus in respect of perceptual issues."

Chowning et al. (1984, pp.17-19) and Schloss (1985, p.90) perform peak-picking on a smoothed IOI histogram, and keep the highest peak, qualifying it as the "important duration." Likewise, Rosenthal (1992, p.41) takes the maximum peak as being the tactus, using a peak-picking algorithm with a bias towards smaller IOIs. In (Foote and Uchihashi, 2001) "beat spectrum," the pulse period is determined as the maximal peak, also by peak-picking. In (Brown, 1993), the pulse of interest is the measure. All the peaks in the ACF are detected and the measure period is taken from the peak whose height is greater than those of all previous peaks and all subsequent peaks up to twice its period. Goto and Muraoka (1995) consider the maximum peak in a *restricted* region (between 61 and 120 BPM). Multiplying the periodicity

function by a tempo preference (probability) distribution (see page 24) may also help
the selection among prominent peaks.

We have seen that in some cases (e.g. comb filterbanks), the computation of the
periodicity function itself accounts implicitly for the fact that the periods of metrical
levels are related with simple integer ratios: major periodicities contribute to the
raising of several peak magnitudes at rationally-related periods. Some systems also
account *explicitly* for the constraints posed by the metrical hierarchy in the parsing
of periodicity function peaks (e.g. the period of any metrical level is an integer multiple of the smallest level period). For instance, in addition to their salience, peaks
can be selected by their alignment with a periodic grid (e.g. computing the autocorrelation function of the periodicity function itself, (Gouyon and Herrera, 2003a)).
Another way to account for the constraints posed by the metrical hierarchy is to
seek periodicities in feature lists computed at the scale of a lower metrical level as
the tatum (Gouyon and Herrera, 2003a; Uhle et al., 2004), instead of frame features.
Dixon et al. (2003) collect peaks from several periodicity functions, consider exhaustively all pairs of peaks as possible tactus/measure combinations, and compute the fit
of all peaks to each hypothesis (see more details on page 62). Another method is to
express the constraints posed by the metrical hierarchy in a probabilistic framework
(Klapuri et al., 2005) (see more details on page 63). This is somehow complementary
to Desain's "(de)composable theory of rhythm perception" (1992) that highlights the
need to consider events with respect to a rhythmic context. Here, the context amounts
to the notion of metrical hierarchy rather than the notion of temporal expectancy at
a single level.

Some systems postpone the parsing of the periodicity function to the tracking
phase: several prominent periodicity peaks are selected and considered by a pulse
tracking algorithm that will eventually favor one tempo hypothesis over the others.

Some systems (e.g. comb filter, tempogram) compute the pulse phase (hence
all beat positions) jointly with the period. In other cases (e.g. ACF), the computation of the period entails the loss of time localization, and the phase has to be
computed subsequently, either during pulse tracking, e.g. (Dixon, 2001a), or by enumerating possible phases once the period is known, and calculating the best match

(Gouyon et al., 2002), see on page 170.

2.2.3 Pulse tracking

Pulse tracking and pulse induction often occur as complementary processes. Pulse induction models consider short-term timing deviations as noise, assuming a relatively stable tempo, whereas a pulse tracker handles the short-term timing deviations and attempts to determine changes in the pulse period and phase, without assuming that the tempo remains constant. Another difference is that induction models work bottom-up, whereas tracking models tend to follow top-down approaches, for example, driven by the pulse period computed by the pulse induction module.

Pulse tracking is often implemented with online algorithms, making real-time implementations possible. Previous data is used to compute pulse period and phase that are used as predictions propagated onto incoming data, and tracking is then a process of reconciliation between these predictions and the observed data. An important part of this process is entrainment, adapting the pulse period and phase based on the observations, which must find a good balance between *reactiveness* and *inertia*. Reactiveness determines how quickly the system responds to a change, and reflects the importance given to the incoming data, while inertia determines the stability of the system and reflects the importance attached to the context given by past data.

Diverse formalisms and techniques have been used in the design of pulse trackers: rule-based, problem-solving, agents, adaptive oscillators, dynamical systems, Bayesian statistics and particle filtering. The framework of state models is general enough to describe and compare pulse trackers: they can all be defined by

- a set of state variables,

- an initial situation (initial values for these variables),

- observations (incoming data),

- a goal situation (finding the best explanation for the observations),

- a set of actions (adapting the state variables in order to reach the goal situation) and

- methods to discriminate good and bad actions.

In the remainder of this section, we review how diverse models deal with the adaptation of state variables to the observations.

2.2.3.1   Observations and state variables

Observed musical events are usually onset features: onset times, durations (or IOIs) and dynamics. Tracking models follow two different rationales regarding observations: they either consider events *sequentially* (i.e. each incoming event is processed and influences the tracker) or they consider *predicted beat positions* (i.e. only events around predicted beats are processed; others are disregarded). State variables usually account for the pulse period, or tempo, and the pulse phase, expressed as either the current beat position or the first beat. Some models also include other variables, such as the estimated metrical position or a performance measure indicating the tracker's self-evaluation, e.g. Dixon (2001a).

2.2.3.2   Actions

Oscillators   Adaptive oscillators predict the next beat position as the current beat position plus the pulse period, and then choose the closest event to this predicted position and adapt the state variables accordingly (McAuley, 1995; Large and Kolen, 1994). For instance in (Large and Kolen, 1994) a simple oscillator, called the "driven" unit, embodies the period and phase variables and adapts to incoming events emitted by the "driver" unit. Each event from the driver perturbs the phase of the driven unit by an amount determined by the coupling strength, which in turn determines the balance between reactiveness and inertia of the model. The resulting instantaneous period of the driven unit eventually differs slightly from its preferred period without coupling. However, in this "phase-pulling" scheme, if the driver stops (i.e. no more input to the driven unit), then the driven unit recovers its previous instantaneous period. The stability of such a system is function of the driven/driver period ratio and the coupling strength (Large and Kolen (1994) provide insightful diagram illustrations). In order to prevent the driven unit from returning to its former period if the driver stops, a "frequency-locking" procedure is also needed. Phase and frequency locking is achieved by minimizing the gap between the current beat prediction and the subsequent event of the driver, according to the method of gradient descent. If this gap is too big, the new event will not be taken into account. Several coupling values are tested and results are detailed in (Large and Kolen, 1994). Large and Kolen (1994) and McAuley (1995) suggest connecting several oscillators in a network so that they can interact, in order to model several metrical levels jointly, see also (Eck, 2001; Gasser et al., 1999) on this issue.

Rule-based approach   In the rule-based approach, e.g. (Desain and Honing, 1999) and (Longuet-Higgins and Lee, 1982), the state variables are the pulse period and the first and current beats. A set of "if-then" rules adapts these variables as each event is observed, and predicts the next beat. For instance, in (Longuet-Higgins and Lee, 1982), a beat is predicted at the current beat position plus the pulse period, and the pulse period is then adapted by two rules: "conflate" and "stretch." The former achieves a doubling of the pulse period when an onset is observed on the predicted beat, the latter changes the period if an onset is observed before the predicted beat (then the period is set to the distance between this new onset and the penultimate beat). Pulse phase is adapted by the rule "update": if no onset is observed at the predicted beat (nor before it), the first beat is shifted to the current beat and the current beat to the predicted beat (regardless of the fact that there is no onset there). This approach seems biased towards reactiveness rather than inertia.

Multiple hypotheses   In (Dannenberg and Mont-Reynaud, 1987), incoming events are considered sequentially and the pulse period is updated as follows. An integer divisor (or multiple) of the pulse period is assigned to the next observation (e.g. 1, 1/2, 1/3, 2, etc.) as the closest metrical position to the actual event position. The resulting deviation then serves to update the pulse period. This updating mechanism depends also on the event position in the metrical hierarchy: events close to multiples of the expected pulse period have a greater impact on the updating mechanism than other events, e.g. half-periods, see (Dannenberg and Mont-Reynaud, 1987, "Confidence" parameter). Finally, the balance between reactiveness and inertia is explicitly monitored by the "Decay" parameter.

Allen and Dannenberg (1990) propose to add some flexibility to the previous model by fine-tuning the "Decay" and "Confidence" parameters, depending on the musical style. However, observing that this model does not possess the capability to *recover after an error*, they introduce the notion of concurrent hypotheses, where a hypothesis is a *sequence of states*. Incoming events are also considered sequentially in this model, but the system does not commit to a decision at each observation. Rather, the evolutions of several concurrent hypotheses are evaluated with some delay with respect to real-time, so that decisions are not taken on the basis of a given state, but on the basis of a sequence of states. In addition to the period and phase variables, a metrical position and a "credibility" (performance measure) are also state variables. In this framework, the number of hypotheses increases with each observation, resulting in a search tree. The tree is pruned to an acceptable size by discarding some hypotheses based on heuristics which implement simple aspects of musical knowledge (e.g. "quarter-notes must start on the downbeat or the upbeat"). Other techniques to reduce the number of hypotheses are by using best-first search,[9] discarding hypotheses which duplicate the current state of other hypotheses, and limiting the number of likely metrical positions. Temperley and Sleator (1999) use dynamic programming to search the solution space of possible mappings of events to a pulse, where the search is guided by a set of preference rules based on GTTM.

Agents   Dixon (2001a) presents another multiple hypothesis search approach, using an agent paradigm, where each agent has a state (state variables are period and phase of a pulse) and a history ("the sequence of beat times selected to date by the agent"). These agents are comparable to the hypotheses of Allen and Dannenberg (1990), except that observations are only processed if they occur around the predicted

---

[9]expanding the best hypotheses first: setting a maximum number of concurrent hypotheses and, when this number is reached, use a heuristic function to determine how well each hypothesis performed so far (e.g. sum of state credibilities in that hypothesis), and keep the best ones.

beats, i.e. "within a window whose width depends on the pulse period."

Dynamical system Cemgil et al. (2001) address pulse tracking through the use of a dynamical system, a "metronome model" that updates state variables at each inferred beat. The system is defined with two hidden state variables: the period and the phase of a metronome. Transition from one metronome beat to the next is modeled by a simple set of state equations. The model is fully determined if the initial state variables are given. To this deterministic model, they add a noise term (a Gaussian random vector whose covariance matrix will be estimated through a training phase) that models the likely tempo variations. Observations to the dynamical system ("noisy metronome beats") are given by the computation of a "tempogram" from incoming onsets. The hidden state variables are estimated by means of a Kalman filter and extensions to the Kalman filter are proposed.

2.2.3.3   Tracking as repeated induction

Some systems address pulse tracking by repeated pulse induction, e.g. (Chung, 1989; Scheirer, 1998; Foote and Uchihashi, 2001; Goto, 2001; Klapuri et al., 2005). A pulse is induced on a short analysis window (usually around 5 s of data), then the window is shifted in time to include the next event and another induction step takes place. (If the feature list consists of frame features, the hop size is constant.) In this framework, observations to the tracking process are no longer events as used above, but the period and phase of a pulse. Determining the tempo evolution is then reduced to connecting the observations at each step.

In addition to computational overload (in comparison to other tracking strategies), one problem that arises with this approach to tracking is the lack of continuity between successive observations. Each induction stage produces pulse period and phase estimations that are usually independent. A subsequent process must connect successive estimations. A continuity constraint is implicitly present in the fact that the analysis hop size is usually much smaller than the window size but this results in a strong bias towards inertia rather than reactiveness and an impossibility to model sharp tempo changes.

If each induction step yields several pulse period and phase hypotheses, finding the final tempo curve and beat locations sums up to finding the "best" path that connects successive hypotheses, e.g. by dynamic programming (Laroche, 2003). "Best" can be formalized, here again, by costs assigned to the several ways of adapting state variables (i.e. pulse periods and phases and measures of self-evaluation). For Goto (2001) and Laroche (2003), this entails continuity and non-syncopation constraints.

2.2.4   Time signature determination

Few algorithms for time signature determination exist. The simplest approach is based on parsing the peaks of the periodicity function to find two significant peaks, which correspond respectively to a fast pulse, the time signature denominator, and a slower pulse, the numerator (Brown, 1993). The ratio between the pulse periods defines the time signature.

Another approach is to consider all pairs of peaks as possible tactus/measure combinations, and compute the fit of all periodicity peaks to each hypothesis, using a weighted sum, where the weights represent the likelihood of each metrical unit appearing as a strong periodicity, given the meter (Dixon et al., 2003).

The time signature is implicitly calculated by systems that induce a complete metrical structure, e.g., (Temperley and Sleator, 1999).

Another strategy is to break the problem into several stages: the determination of the time signature denominator (e.g. by tempo induction and tracking), the segmentation of the musical data with respect to this pulse, the definition of features at this temporal scope and subsequently the detection of periodicities in feature lists (Meudic, 2002; Gouyon and Herrera, 2003b). Goto and Muraoka (1999) detect chord changes as indicators of higher level metrical boundaries such as bar lines; however their work is restricted to music with a $\frac{4}{4}$ time signature.

2.2.5   Rhythm parsing

Rhythm parsing, or quantization, is the process aiming at making a list of note onsets fit into Western music notation, see page 22. It can be seen as a by-product of the induction of several metrical levels, which together define the metrical structure, e.g., for Chung (1989, p.21), "obtaining the correct metric and rhythmic interpretation are part of the same process." Onset times of a given sequence can be parsed by assigning each onset (independently of its neighbors) to the closest element in this metrical structure. The weaknesses of such an approach are that it fails to account for musical context (e.g. a triplet note is usually followed by 2 more) and tempo changes.

Models by Desain and Honing (1989) and Cemgil et al. (2000) do account for musical context and possible distortions of the metrical structure. However such distortions would in turn be easier to determine if the quantized durations were known (Allen and Dannenberg, 1990). Therefore, rhythm parsing is often considered as a process *simultaneous* with beat tracking,[10] rather than subsequent to it (hence the bi-directional arrow between these two modules in Figure 2.4 on page 32).

Here also, as for beat tracking on page 51, the joint estimation of tempo and quantized durations can be seen as a process of reconciliation between predicted values for state variables and incoming observations. The main difference with beat tracking lies in the fact that the state variables explicitly specify the interdependency between tempo and quantized durations.

For instance, in (Rosenthal, 1992) the state variables account for three metrical levels simultaneously and observations are not defined by events considered sequentially, but only by events which are close to beats at one of the metrical levels. The pruning techniques are comparable to those used by Allen and Dannenberg (1990), but adapted to the fact that states (and therefore hypotheses) are more complex (Rosenthal, 1992, pp.57-68).

In (Raphael, 2002) and (Cemgil and Kappen, 2003), events are considered sequentially. Concurrent hypotheses are expressed as posterior probabilities of a probabilistic model whose hidden layers (i.e. state variables) account for score notation and ideal timing in addition to tempo. They implement different strategies for parsing the tree of hypotheses and keeping it from growing exponentially. For instance, particle filters are suitable (see also (Hainsworth and Macleod, 2003, 2004), for a similar approach

---

[10]Hence the expression "rhythm tracking," occasionally found in the literature.

using audio data). Temperley and Sleator (1999) also process events sequentially, using dynamic programming and a simple set of preference rules to infer up to 5 metrical levels.

Thornburg (2001a) also follows the same rationale, however he includes audio segmentation (onset detection) as a third interdependent process (Thornburg, 2001b), rather than a preprocessing step before rhythm parsing and beat tracking. He argues that these tasks should be considered jointly: polyphonic audio segmentation is necessary to provide data to the rhythm tracker, but rhythm tracking should also orient (i.e. provide prior probabilities to) the segmentation task. This helps to ensure robustness against spurious onsets, which are a common problem in polyphonic audio segmentation. The systems based on MIDI input (Temperley and Sleator, 1999; Raphael, 2002; Cemgil and Kappen, 2003) account inherently for noise in onset timing, but not for spurious onsets.

2.2.6   Systematic deviation estimation

In the pulse induction process, short term-timing deviations can be "smoothed out" (see page 45) or cautiously handled so as to derive patterns of short-term timing deviations such as swing. Foote and Uchihashi (2001) suggest that swing could be measured by inspection of a periodicity function (there, the "beat spectrum") within the pulse induction process. This is illustrated by the positions of secondary peaks with respect to some higher ones in (Foote and Uchihashi, 2001, Figure 3) but they do not suggest any extraction procedure. Another problem is that periodicity functions do not distinguish the order of events, e.g., the difference between a long-short pattern and a short-long pattern, which is critical.

Laroche (2001) proposes to estimate the swing jointly with tempo and beats at the half-note level, assuming constant tempo. The procedure is conceptually similar to pulse induction using a pulse matching function (see page 45), but enumerating all possible pulse periods and phases, like Cemgil et al.'s Tempogram (see page 44), and searching for the one which best matches the onsets. The number of candidate pulses (the search space) is in fact even larger, as tracks have a third parameter to be estimated (the swing) in addition to the tempo and phase parameters. In this case the pulses are no longer isochronous, but correspond to the long-short timing pattern that we wish to find in the data. The amount of deviation from an isochronous track defines the swing ratio. Gouyon et al. (2003) estimate swing ratio in a comparable fashion, see on page 188.

2.2.7   Rhythm pattern determination

The previous section reported on the relevance of systematic short-term timing patterns. In addition, repetitive patterns covering *longer* temporal scopes can also be characteristic of some music styles. For instance, many electronic synthesizers feature templates of prototypical patterns such as Waltz, Cha Cha and the like. The length of such patterns is typically one bar, or a couple or them. Few algorithms have been proposed for the automatic extraction of rhythmic patterns; they usually require the knowledge (or previous extraction) of part of the metrical structure, typically the tactus and measure beats (Dixon et al., 2004).

2.2.8   Periodicity features

Other rhythmic features, with a musical meaning less explicit than e.g. the tempo or the swing, have recently been advocated, in particular in the context of designing rhythm similarity distances. Most of the time, these features are derived from a parameterization of a periodicity function, as e.g. the salience of several prominent peaks (Gouyon et al., 2004a), their positions (Tzanetakis and Cook, 2002; Dixon et al., 2003), selected statistics (high-order moments, flatness, etc.) of the periodicity function considered as a probability density function (Gouyon et al., 2004a) or simply the whole periodicity function itself (Foote et al., 2002).

2.3   Evaluation of tempo induction systems

Among the diverse aspects of automatic rhythm description, most of the effort has been dedicated to the tempo induction and beat tracking tasks. Section 2.2 provided a

review of the diverse formalisms that have been used to implement computer systems
performing these tasks. Within any computational modeling endeavor, systematic
evaluations of competing models is highly desirable. They require:

- Reference examples of correct analyses, that is, large and publicly available
  annotated data sets

- An agreement on the manner of *representing* and *annotating* relevant information about this data

- *Agreed evaluation metrics*

Such evaluations have received little attention in pulse induction and tracking. Early
models usually did not present quantitative evaluation of the proposed models, and
only recently have researchers begun to report on the performance of their systems,
but they meet with the following difficulties.

First of all, even if a number of papers propose evaluation methodologies, no
consensus has been reached on how to evaluate algorithms, because of the diversity of input and output data representations as well as the diversity of applications (Gouyon and Meudic, 2003). For instance, Temperley (2004) convincingly highlights shortcomings of metrics proposed by Goto and Muraoka (1997) and Cemgil et al.
(2001), and proposes an evaluation method that seems suitable for systems processing MIDI input. However, as this metrics is based on a note-by-note evaluation
(not beat-by-beat), in order for it to be useful for acoustic signal inputs, it would
require complete transcriptions of these signals, an unrealistic requirement from the
point of view of manual annotation, and well beyond the scope of the pulse induction
algorithms themselves.

Secondly, the evaluation data sets used by many researchers are usually private
and of relatively small size, which makes it difficult to compare one system with
another. Some efforts have been made to make data public. For example, a collection
of score-matched MIDI performance data is available from the Music, Mind and
Machine Group of the University of Nijmegen[11] (around 200 performances of a couple

---

[11]http://www.nici.kun.nl/mmm/archives/

of Beatles songs by 12 pianists performed in several tempo conditions). Results on this data set were reported by Cemgil et al. (2001) and Dixon (2001b), and the latter argued that more challenging data was needed. Also, Temperley (2004) provides a publicly available data set[12] of 46 pieces with metronomical timing and 16 performed pieces, all taken from the common-practice Western repertoire. However, in both cases, the data sets are only suitable for evaluating systems dealing with MIDI input, and not acoustic signal input.

Finally, there are many models, but few open source implementations, and few models are described completely enough in order to reimplement them.

As a first step towards more systematic evaluations and comparisons, we organized a tempo induction contest during the International Conference on Music Information Retrieval (ISMIR 2004) held at the University Pompeu Fabra in Barcelona in October 2004.[13] The task was restricted to the induction of tempo as a scalar, in beats per minute (and not the individual beat positions or any other rhythmic description). Researchers were encouraged to participate by several means, the respondents agreed upon an evaluation benchmark for the competition. In the remainder of this section, we report on the results of this competition.

2.3.1 Algorithms

Twelve algorithms entered the contest, 11 were submitted by 6 different research teams, and one open-source algorithm (GPL-licensed) was downloaded from the web. The contest organizer did not compete. One entrant chose not to participate in the analysis of the competition results, so we report here on 11 algorithms. Algorithms were submitted in various formats: the open-source entries were submitted as C, C++ or Matlab source code, and the others as Windows or GNU/Linux binaries or Matlab pre-parsed pseudocode files. All of the algorithms are based on a common general scheme: a *feature list creation* block, that parses the audio data into a temporal series of features which convey the predominant rhythmic information to the following *pulse induction* block (see Figure 2.4 and Section 2.2 for more details). Some algorithms also

---

[12]`ftp://ftp.cs.cmu.edu/usr/ftp/usr/sleator/melisma2003`
[13]The conference webpage is `http://ismir2004.ismir.net/`

implement a *beat tracking* block. However, as the contest did not address the issues
of tracking tempo changes and determining beat positions, the submitted algorithms
either bypassed this block or added a subsequent back-end for the purpose of the
contest, i.e. a parsing of the beat positions into a global tempo estimation.

2.3.1.1   Algorithms by Alonso et al. (2004)

Miguel Alonso from the École Nationale Supérieure des Télécommunications (ENST)
in Paris submitted two algorithms, referred to as AlonsoACF and AlonsoSP, which
were submitted in the form of p-files, i.e. Matlab pre-parsed pseudocode files (source
code is not visible).

Both methods are based on the same front-end that extracts onsets of notes.
A time-frequency representation of the audio signal is calculated, and the rate of
change of the spectral energy content is found by filtering this representation with a
differentiator FIR filter. The positive contributions of each spectral line are summed
in the frequency domain and an onset train is obtained.

The difference between the systems is found in the pulse induction block. The
first method is based on the autocorrelation of the onset train, while the latter uses
the spectral product. Both algorithms are described in detail in (Alonso et al., 2004).
These systems were originally conceived for beat tracking, but the tracking part was
disabled in the versions submitted to the contest.

2.3.1.2   Algorithms by Dixon (2001a) and Dixon et al. (2003)

Simon Dixon from the Austrian Research Institute for Artificial Intelligence (ÖFAI)
in Vienna submitted three entries to the contest: DixonI, DixonT and DixonACF.

The first two are GNU/Linux binaries based on the beat tracking system BeatRoot
detailed in (Dixon, 2001a).[14] They are both based on a simple energy-based onset
detector followed by an inter-onset interval (IOI) clustering algorithm. DixonI selects
a tempo based on the "best" cluster, where the clusters are assessed by the number of
IOIs they contain, the amplitude of the corresponding notes, and the support of other

---

[14]BeatRoot is available as GPL code at http://www.oefai.at/~simon/beatroot/index.html

clusters related by simple integer ratios. DixonT selects several prominent clusters as tempo hypotheses, performs beat tracking based on these hypotheses, and outputs the mean of the inter-beat intervals (IBI) from the best beat tracking solution as the final estimate of tempo.

DixonACF (Matlab source code) is described in (Dixon et al., 2003). This algorithm splits the signal into 8 frequency bands, and then smooths, downsamples and performs autocorrelation on each of the frequency bands. From each band, the 3 highest peaks (excluding the zero-lag peak) of the autocorrelation function are collected. The tempo is derived from this list of peaks by taking into account influential schemes between metrical levels: the algorithm considers exhaustively all pairs of peaks as possible tactus/measure combinations, and computes the fit of all periodicity peaks to each hypothesis, using a weighted sum, where the weights represent the likelihood of each metrical unit appearing as a strong periodicity, given the meter (Dixon et al., 2003).

2.3.1.3 Algorithm by Klapuri et al. (2005)

Anssi Klapuri from the Tampere University of Technology submitted one algorithm as a GNU/Linux binary, referred to as Klapuri.

An important aspect of this algorithm lies in the feature list creation block: the differentials of the loudness in 36 frequency subbands are combined into 4 "accent bands", measuring the "degree of musical accentuation as a function of time." The goal in this procedure is to account for subtle energy changes that might occur in narrow frequency subbands (e.g. harmonic or melodic changes) as well as wide-band energy changes (e.g. drum occurrences). The pulse induction block implements a bank of comb filters comparable to that proposed by Scheirer (1998) (see below).

Another particularity of this algorithm is the joint determination of three metrical levels (the tatum, the tactus and the measure) through probabilistic modeling of their relationships and temporal evolutions. After computing the tactus beats of the whole test excerpt, the tempo was computed as the median of the IBIs of the excerpt's latter half. See (Klapuri et al., 2005) for a complete description of the algorithm.

2.3.1.4   Algorithm by Scheirer (1998)

The source code of Eric Scheirer's algorithm (formerly MIT Media Lab) was down-loaded from the web (`http://sound.media.mit.edu/~eds/beat/tapping.tar.gz`), it was ported to GNU/Linux (it is referred to as Scheirer).

Scheirer (1998) performs pulse induction with a comb filterbank on a regularly-sampled signal amplitude envelope separately on 6 frequency bands which are then combined *after* periodicity detection. The output of the algorithm is a set of beats rather than an overall tempo estimate, so we added a small back-end to the code that outputs the state of the filterbank after the analysis of the whole sound file. Then the tempo is taken to be the resonance frequency of the filter with the highest instantaneous energy after the whole analysis. The choice of this particular back-end was based on the observation that this algorithm provides more reliable estimates after some processing of the sound file than at the beginning. However, note that other methods could also be considered, as for instance, the total number of beats divided by the duration, or the mean of the IBIs.

2.3.1.5   Algorithms by Tzanetakis and Cook (2002)

George Tzanetakis from Victoria University submitted 3 entries: TzanetakisH, Tzane-takisMS and TzanetakisMM (standing respectively for "Histogram", "MedianSum" and "MedianMultiband"). GNU/Linux binaries were compiled from the source code available on the SourceForge web.[15]

All the three methods are based on the wavelet front-end described in (Tzanetakis and Cook, 2002). The signal is segmented in time into 3 s analysis windows (with an overlap of 1.5 s). In each window, the signal is decomposed with the help of a wavelet transform into 5 octave-spaced frequency bands, and the amplitude envelope is extracted in each band.

Regarding the pulse induction block, all three methods use autocorrelation, how-ever, they differ in some aspects. The default method (TzanetakisMS) sums the

---

[15]marsyas-0.2 under `http://www.sourceforge.net/projects/marsyas`

diverse subband amplitude envelopes and computes an autocorrelation of the resulting sum. The maximum peak in the autocorrelation (a tempo estimate) is computed on each analysis window and the median of the tempo estimates is chosen as the final tempo. TzanetakisMM makes a separate tempo estimate for each band and each analysis window, and then selects the median. TzanetakisH sums the subband amplitude envelopes, computes an autocorrelation of the resulting sum, selects several autocorrelation peaks and accumulates them in a histogram which summarizes the peaks of all analysis windows. The tempo is finally set to the highest peak of the histogram.

2.3.1.6 Algorithm by Uhle et al. (2004)

Christian Uhle from Fraunhofer Institute for Digital Media Technology submitted one algorithm as a Windows binary, referred to as Uhle. This algorithm calculates the rates of three pulses (the tatum, the tactus and the measure). The audio signal is segmented into characteristic long-term segments corresponding for example to a verse or a chorus (Foote, 2000). Amplitude envelopes are calculated for logarithmically-spaced frequency bands by means of the Discrete Fourier Transform and smoothed using an FIR low-pass filter. The first-order differential of each band is computed, and subsequently half-wave rectified. They are then summed across all bands to produce an "accent signal." An autocorrelation function is computed for non-overlapping 2.5 s segments inside each long-term segment. The tatum period is estimated from the ACF by means of the two-way mismatch error function (Gouyon et al., 2002).

A second periodicity function is derived from the ACF in order to highlight the saliences of periodicities only at integer multiples of the tatum, among which are the measure and the tactus. Instead of simply downsampling the ACF by a factor equal to the tatum, this function is computed by selecting local maxima around the positions of integer multiples of the tatum. This method is comparable to that proposed by Gouyon and Herrera (2003a), where an ACF is computed on low-level features computed at the temporal scale of the tatum (instead of the frame). An important difference is that the phase of the tatum is not required in the method proposed by Uhle.

This second periodicity function is compared (i.e. correlated) with a number of pre-defined metrical templates, which characterize musical knowledge of different meters. The current implementation has 17 templates. The most highly correlated template determines the value of the segment's tempo (and incidentally measure). Tempi are accumulated in a weighted histogram and the maximum yields the basic tempo of the piece. See (Uhle et al., 2004) for more details.

2.3.2   Experimental framework

2.3.2.1   Infrastructure

Two computers were used: AlonsoACF, AlonsoSP and Uhle were run on Windows OS (XP Professional edition 2002, version 5.1.2600), the rest on GNU/Linux OS (Debian Sarge), both 1.6 GHz, with 512 MB RAM. The evaluation framework was designed as a set of Matlab[16] (version 6.1, Release 12.1 on GNU/Linux and version 6.5, Release 13 on Windows), perl, shell and dos scripts. For a robustness test (see below), several types of distortion were applied to the signal using the programs Sox and Matlab. However, it was ensured that the tempo was still clearly perceivable even in the cases of severe degradation of signal quality. All of the test scripts are available from the contest webpage.

2.3.2.2   Data

No training data was provided. However, some preparatory data (7 pieces and corresponding tempo values) was given to the participants in order to compare whether algorithms yield the same output when run in participants' labs and on the machines used for the contest, and to check proper formatting of algorithm input and output.[17]

The test data consisted of 3199 tempo-annotated pieces in 3 data sets as described below. The pieces range from 2 to 30 seconds, and from 24 BPM[18] to 242 BPM. Figure 2.6 illustrates the distribution of test excerpts along the tempo axis (these

---

[16]http://www.mathworks.com

[17]This was not considered as "training data" as it would not be possible to properly train a system with so few pieces and test it on a test set more than 400 times greater.

[18]Note however that only 15 excerpts have a tempo less than 50 BPM

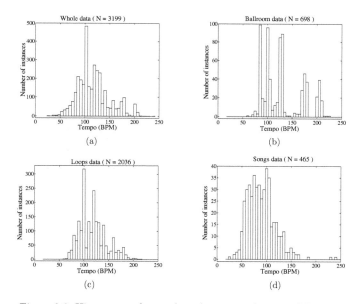

Figure 2.6: Histograms of ground-truth tempo values in 5 BPM steps

tempo statistics are all available in text format on the contest webpage and may
be used e.g. for setting prior probabilities in Bayesian approaches). They all have
approximately constant tempi, and the format is the same for all: mono, linear PCM,
44100 Hz sampling frequency, 16 bit resolution. The total duration of the test set
is approximately 45140 s (i.e. around 12 h 36 min). This data was not available to
participants before the competition. Part of the data has now been made available
on the contest webpage in order to stimulate further research.

Loops   Many sound libraries are made up of short "loops" to be used in DJ sessions,
or for home recording needs. The loops used here were originally in MP3 format with
a relatively low sound quality, they come from different sound library retailers and are
courtesy of the Tape Gallery.[19] It is usual that tempo in BPM (and additional metadata) are sold together with sound files. These annotations were not double-checked.
We do not distribute these loops for copyright reasons, however, an exhaustive list of
loops and corresponding tempo is available on the contest webpage.[20] One can search
by name for, listen to MP3 versions of and buy high audio quality versions of any of
these loops from the webpage of the Tape Gallery.

A loop is often used as a basic short "kernel" to be looped in a composition, that
is, to generate a long audio file by several concatenations. However, the samples used
in the analysis were not looped.

- Total number of pieces: 2036

- Duration: a few bars

- Total duration: around 15170 s

- Tempo range: between 60 and 215 BPM, see Figure 2.6(c)

- Genres: Electronic, Rock, House, Ambient, Techno.

---

[19]http://www.sound-effects-library.com/
[20]http://www.iua.upf.es/mtg/ismir2004/contest/tempoContest/node4.html

Ballroom BallroomDancers.com[21] provides information on ballroom dancing (online lessons, etc.). Some characteristic excerpts of many dance styles are provided in the low sound quality Real Audio format, with a compression factor of almost 22 with respect to the common 44.1 kHz 16 bits mono WAV format, labeled with a tempo value. Tempo values were double-checked by Simon Dixon.

Data and annotations are available on the contest webpage.

- Total number of pieces: 698

- Duration: around 30 s

- Total duration: around 20940 s

- Genres: see style distribution in Table 2.1.

- Tempo range: between 60 and 224 BPM, see Figure 2.6(b)

Style	# pieces
Cha Cha	111
Jive	60
Quickstep	82
Rumba	98
Samba	86
Tango	86
Viennese Waltz	65
Slow Waltz	110

Table 2.1: Style distribution of the ballroom dance music excerpts

Song excerpts A professional musician placed tactus beats on several song excerpts. (These beats were cross-checked by the author of this book). The ground-truth tempo was computed as the median of the IBIs; other methods could also be considered, as for instance, the total number of beats divided by the duration.

Data and annotations are available on the contest webpage.

---

[21]http://www.ballroomdancers.com/

- Total number of pieces: 465

- Duration: around 20 s

- Total duration: around 9300 s

- Genres: see distribution in Table 2.2.

- Tempo range: between 24 and 242 BPM, see Figure 2.6(d)

Genre	# pieces
Rock	68
Classical	70
Electronica	59
Latin	44
Samba	42
Jazz	12
AfroBeat	3
Flamenco	13
Balkan and Greek	144

Table 2.2: Genre distribution of the song excerpts

2.3.2.3   Evaluation methods

Two evaluation metrics were agreed for the contest:

- *Accuracy 1*: The percentage of tempo estimates within 4% (the *precision window*) of the ground-truth tempo.

- *Accuracy 2*: The percentage of tempo estimates within 4% of either the ground-truth tempo, or half, double, three times, or one third of the ground-truth tempo.

The latter evaluation metrics was motivated by the fact that the ground-truth we use for evaluation does not necessarily represent the metrical level that the majority of human listeners would choose. However, we assume that discrepancies between ground-truth tempo and human perception correspond to a focus on a different metrical level, i.e., a ratio of 2 or $\frac{1}{2}$ for duple meter music and a ratio of 3 or $\frac{1}{3}$ for triple meter music. This assumption is ubiquitous in all previous evaluation attempts; see on page 70 for further discussion. As we discuss in further details on pages 70 and 75, the width of the precision window is not a crucial factor.

In addition, the robustness of algorithms to sound distortion was evaluated on a part of the test data: the 465 Songs excerpts. These pieces were distorted by several processes: downsampling/resampling, GSM encoding/decoding, filtering, volume change and addition of reverberation and white noise (with a signal-to-noise ratio of 20 dB). The script is available on the contest webpage and in the appendix on page 253.

2.3.3 Results

2.3.3.1 Accuracy measures and robustness to noise

	Ballroom		Loops		Songs	
	acc1	acc2	acc1	acc2	acc1	acc2
AlonsoACF	29.8	64.76	28.09	55.45	23.44	58.28
AlonsoSP	34.1	69.48	36.79	70.14	37.42	68.6
DixonACF	43.12	86.96	42.34	**81.93**	16.99	76.99
DixonI	25.79	65.76	34.53	78.14	28.6	65.58
DixonT	39.4	79.8	23.82	73.67	19.35	68.82
Klapuri	**63.18**	**90.97**	**70.71**	81.57	**58.49**	**91.18**
Scheirer	51.86	75.07	33.06	65.37	37.85	69.46
TzanetakisH	23.78	66.05	26.62	52.36	21.29	47.74
TzanetakisMM	32.81	52.29	32.81	52.41	18.71	41.08
TzanetakisMS	33.81	63.18	31.19	53.59	27.53	52.47
Uhle	56.45	81.09	52.16	75.39	41.94	71.83

Table 2.3: Algorithm accuracies 1 and 2 (in %) on the Ballroom data set, the Loops data set and the Songs data set

Figure 2.7 presents the results, on the whole dataset and on each individual subset, for each algorithm, ordered alphabetically: A1 is AlonsoACF, A2 is AlonsoSP, D1

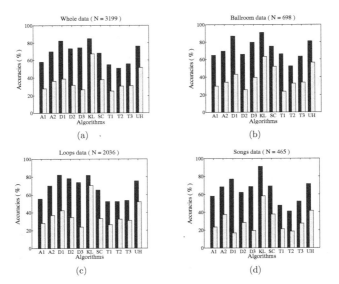

Figure 2.7: Algorithm accuracies 1 (light) and 2 (dark) on the whole data set –2.7(a)–, the Ballroom data set –2.7(b)–, the Loops data set –2.7(c)– and the Songs data set –2.7(d).

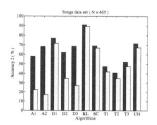

Figure 2.8: Effect of distortions on accuracy 2, dark bars for clean data, light bars for distorted data.

is DixonACF, D2 is DixonI, D3 is DixonT, KL is Klapuri, SC is Scheirer, T1 is Tzanetakis H, T2 is TzanetakisMM, T3 is TzanetakisMS and UH is Uhle. For each algorithm, accuracy 1 and 2 are given, in light and dark shadings, respectively, for the whole data set and each of the 3 subsets. Table 2.3 provides the exact numbers for the individual data sets.

Figure 2.8 illustrates the loss of accuracy for each algorithm when distortion was applied to the Songs data set as detailed above. Clearly, algorithms AlonsoACF, AlonsoSP, DixonI and DixonT suffer more from distortions than other algorithms.

The performance measures accuracy 1 and 2 were the criteria used to determine the contest winner. Klapuri outperforms the other algorithms with respect to both measures on the whole data set: respectively 67.29% and 85.01%. As can be seen in Figure 2.7, it also outperforms the others on most of the individual data sets: {70.71%, 81.57%}, {63.18%, 90.97%} and {58.49%, 91.18%} on the Loops, Ballroom and Songs data sets, respectively. It is also the best algorithm with respect to noise robustness (loss of 1.72 percentage points in accuracy 2, see Figure 2.8).

Statistical significance One must keep in mind that, because of the restriction to a specific data set, the numbers reported in Figure 2.7 are just *estimates* of the true (but unknown) algorithm accuracies. Therefore, in addition to providing success rates for each algorithm, it is important to consider whether the observed differences in performance are statistically significant or arise by chance.

Different statistical tests can be used to compare algorithms based on their respective predictive accuracy: e.g. a test for the difference of two proportions, Student's *t* test, McNemar's test, cross-validation paired differences *t* tests (Dietterich, 1998). Choosing the appropriate test to a given problem depends on the suitability of several assumptions, among them independence of algorithm accuracies (i.e. accuracies on test items are independent for algorithm A and algorithm B) and error independence between items (i.e. errors made by an algorithm on two separate test items are independent).[22]

In our problem, algorithms are all tested on the same pieces, therefore we cannot assume that, for a given piece, the failures of different algorithms are independent. On the other hand, it seems reasonable to assume that errors made by a specific algorithm on different pieces are independent. As mentioned in (Gillick and Cox, 1989, Paragraph 3.2) and (Dietterich, 1998, Question 3), McNemar's test is appropriate to this kind of problem.

McNemar's statistical test tests the hypothesis that the fact that algorithm A classifies an item correctly while algorithm B classifies it incorrectly is equally likely as the opposite (algorithm B classifies an item correctly while algorithm A classifies it incorrectly). In other words it tests the fact that given only one algorithm makes an error, it is equally likely to be either one (this is the "null hypothesis"). Given a threshold for statistical significance (usually 0.01 or 0.05) the null hypothesis is tested by applying a two-tailed test with a Normal distribution (see (Gillick and Cox, 1989) for more details).

According to this statistical test, the observed difference (of around 1%) in accuracy 1 on the whole data set (see Figure 2.7(a)) between AlonsoACF and DixonT would arise by chance on 19% of occasions, this difference is therefore not statistically significant (considering a p-value of 0.01 as the threshold for statistical significance), and it is better to conclude that both algorithm performances are comparable. Similarly, observed performance differences between AlonsoSP and DixonACF (less than 3%), AlonsoSP and Scheirer (less than 2%), DixonACF and Scheirer (1%), DixonI and

---

[22]which does not mean that algorithm accuracy would be independent of the test set

TzanetakisMM (1%), DixonI and TzanetakisMS (less than 2%), DixonT and TzanetakisH (less than 2%) and TzanetakisMM and TzanetakisMS (less than 1%) are not
statistically significant, setting the threshold for significance to a p-value of 0.01.[23]
The differences between all remaining pairs of algorithms are representative of genuine
performance differences.

Regarding accuracy 2, solely the differences between AlonsoACF and TzanetakisMS
(less than 3%), AlonsoSP and Scheirer (less than 2%), DixonI and DixonT (less than
1%), DixonT and Uhle (less than 2%) and TzanetakisH and TzanetakisMS (less than
1%) are not significant.[24] The differences between all remaining pairs of algorithms
are statistically significant.

Computation time Another interesting aspect of the algorithms is the computational resources they require. It can be expressed as processing time divided by
excerpt length:[25] DixonI takes approximately 0.02 times the excerpt length to estimate its tempo, DixonT, Uhle, AlonsoSP and AlonsoACF approximately 0.1, Scheirer
approximately 0.4, Klapuri approximately 0.5, DixonACF approximately 1 and TzanetakisH, TzanetakisMM and TzanetakisMS approximately 2. (Note that the participants
were not instructed to optimize computational efficiency when submitting the algorithms and using different operating systems and versions of Matlab may have an
influence on computation time.)

To facilitate comparison of other algorithms with those of the contest, detailed
results on each dataset are available on the contest webpage.[26]

2.3.3.2 Error analysis

Accuracy vs precision window width Figure 2.9 plots the relationship of algorithm performance to precision window width. The choice of 4% precision in accuracies 1 and 2 is somewhat arbitrary. In the literature, other values have been
advocated; for instance, Klapuri et al. (2005) propose a precision of 17.5% for IBIs,

---

[23]They correspond respectively to P-values of 0.02, 0.16, 0.46, 0.25, 0.5, 0.13 and 0.5.
[24]P-values of 0.03, 0.09, 0.18, 0.03 and 0.26
[25]Algorithm computation times are approximately proportional to excerpt length.
[26]http://www.iua.upf.es/mtg/ismir2004/contest/tempoContest/Results.htm

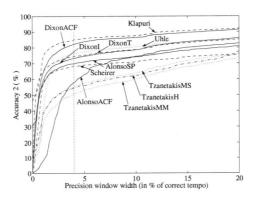

Figure 2.9: Accuracy 2 vs precision window width, full data set

however they focus on *consecutive* IBIs rather than on global tempo and deal with excerpts with varying tempo. The amount of tempo variation in the data is an important factor to consider in setting the precision. Since we are dealing with basically constant-tempo data, a small precision window seems appropriate.

Tendencies towards integer ratio errors   Figure 2.10 shows the type of errors made by the contest winner (Klapuri). Figure 2.11 shows the same information plotted against tempo. One can see on the one hand that the most common errors are doubling and halving of tempo, and on the other hand that it shows a "moderate tempo tendency", i.e. a tendency to estimate half the tempo for fast pieces and double for slow pieces. We remark also that it estimates incorrectly (with respect to accuracy 1) all pieces whose tempi lie outside the rough limits of 60 to 160 BPM. This is due to the explicit modeling of a prior probability function for the tempo (Klapuri et al., 2005; Parncutt, 1994).

Regarding the other algorithms, inspection of their error histograms also shows clearly that, as expected, halving and doubling of tempo are the most common errors. On the other hand, Klapuri seems to be the only algorithm that clearly shows a moderate tempo tendency. With the exception of Klapuri and Scheirer, all algorithms tend

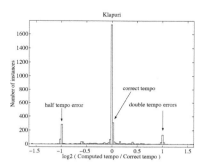

Figure 2.10:   Histogram of ratio between estimated tempo and correct tempo for Klapuri, full test data set

to "tap too fast" rather than too slow. For instance, as can be seen in Figures 2.12(a) and 2.12(b), DixonT has a very clear tendency towards faster metrical levels.

Other typical error factors are $\frac{4}{3}$ and $\frac{2}{3}$, as seen, for example, in the peaks around -0.58 and 0.41 on the (logarithmically-scaled) X-axis of Figures 2.12(a), 2.12(b) and 2.12(c). An error of $\frac{4}{3}$ in the tempo estimation represents an error of $\frac{3}{4}$ in the IBI, that is, a focus on e.g. the dotted quarter-note instead of the half-note, while a tempo error of $\frac{2}{3}$ represents a focus on e.g. the dotted-quarter note instead of the quarter-note.

Algorithms also sometimes estimate $\frac{1}{3}$ of the correct tempo. See, for instance, the peak around -1.58 in Figures 2.12(c) and 2.12(a). This error factor, as well as 3 and $\frac{3}{2}$, are typical of triple and compound meter pieces (e.g. Waltz in the Ballroom data set). We found relatively few of these errors, presumably because relatively few such pieces are present in the test data set.

Algorithm performance "niches"   It is interesting to consider whether specific algorithms, regardless of their overall performance, show unique performance on some particular data. Indeed, an algorithm which performs worse than other algorithms on many problems, but solves a few problems that no other algorithm solves, would be valuable if these special cases could be identified.

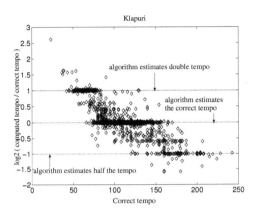

Figure 2.11: Klapuri performance with respect to piece tempi, full test data set

There are 41 pieces (3 ballroom, 35 loops and 3 songs) whose tempi were correctly computed by *all* 11 algorithms. On the other hand, 176 pieces (11 ballroom, 162 loops and 3 songs) were incorrectly processed by *all* algorithms, with respect to both accuracy 1 and accuracy 2. Finally, there are 29 pieces whose tempi were correctly computed by a *single* algorithm. No clear explanations for these cases have been found.

Another way to thoroughly inspect the results is to compare pairs of algorithms. For instance, Figure 2.13 shows a comparison between Klapuri and DixonACF. For each data set, pieces have been ordered with respect to increasing error made by Klapuri, where the error is computed as follows:

$$e = \left| log2 \left(\frac{computedTempo}{correctTempo} \right) \right|$$

The performance of both algorithms is given for each piece, permitting a visual comparison of algorithms on a piece-by-piece basis. Three main trends are apparent: many cases of agreement between the algorithms, for correct and incorrect tempo estimates; cases where one algorithm is correct and the other has a halving or doubling

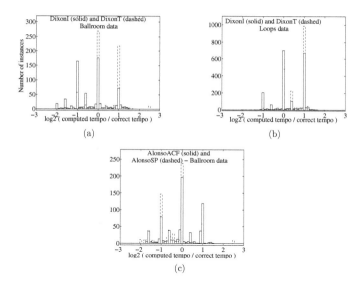

Figure 2.12: Histograms of ratio between estimated tempo and correct tempo for DixonI (solid) and DixonT (dashed), on the Ballroom and Loops data sets —2.12(a) and 2.12(b) respectively— and for AlonsoACF (solid) and AlonsoSP (dashed), on the Ballroom data set —2.12(c).

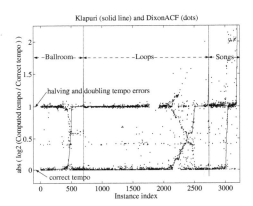

Figure 2.13: Comparison of errors made by Klapuri (solid line) and DixonACF (dots)

error; and cases where both algorithms are incorrect, and one algorithm has double
the tempo of the other.

For example, in the Ballroom data, DixonACF solves quite a few doubling and
halving errors that Klapuri makes (see the cluster of points around the error value of
0 for indexes between 500 and 698), but on this very data it also makes quite a few
doubling or halving errors where Klapuri estimates the correct tempo. This is also
true of the Loops data set (indexes between around 2500 to 2734), but not the Songs
data set, where DixonACF makes many doubling and halving errors (this can also be
seen in Figure 2.7(d), third bar pair from the left). On the other hand, DixonACF
seems to solve some non-integer ratio errors that Klapuri makes, especially in the
Loops data set (indexes between around 2100 and around 2500, where Klapuri's error
on the Y-axis is between 0 and 1). Note that the apparent mirroring of error values
(reflection in the line $y = 0.5$) is an artifact of the representation, which occurs when
one algorithm has a log error e, where $-1 < e < 0$, and the other algorithm has
double this tempo, hence a log error of $e' = e + 1 = 1 - |e|$.

Figures such as Figure 2.13 can be generated for any pair of algorithms. They
show on a piece-by-piece basis which errors an algorithm makes that another one

does not make. We can then track down single files for which a specific algorithm has a particular advantage over another one. Cases where several algorithms make the same error could be used to identify interesting ("pathological") test cases for further investigation, general weaknesses in current tempo induction systems, and errors in annotation. However, in order to draw conclusions about error trends, or alternatively, specific "skills" or "performance niches" of algorithms, much more test data is needed, together with richer metadata. This is left for future work.

2.4 Conclusions

This chapter defined terms and concepts of interest to the whole book in Section 2.1. It then provided in Section 2.2 a qualitative comparison of existing computational systems of rhythm description with respect to the functional units of a general model (Figure 2.4). Finally, it provided a quantitative comparison of a number of pulse induction algorithms in Section 2.3. We now summarize main findings in the field of automatic rhythm description and highlight open issues. We first focus on the first stages of rhythm description, up to the computation of rhythm periodicity functions and pulse induction. Then, on page 86, we address shortly further aspects of rhythm description systems.

2.4.1 Pulse induction

We depicted two procedures for pulse induction: pulse selection and periodicity function computation, and gave examples of various implementations. Computing a periodicity function is usually more powerful than just selecting a pulse. However, there is a trade-off between the length of the induction window and the likelihood that the tempo-stability assumption holds. Using short windows (typical of pulse selection methods) lowers the reliance on a constant tempo but generates less reliable predictors, whereas using longer windows (typical of periodicity computation methods) generates more reliable predictors but only when the tempo remains relatively unchanged over this longer duration.

Quantitative comparisons of tempo induction algorithms are largely absent from the literature. The evaluation reported in Section 2.3 shows that, for music with almost constant tempo, tempo induction is feasible with around 80% accuracy and a relatively good robustness to distortion, if we do not insist on finding a specific metrical level (the most common errors are in the choice of metrical level, the majority of algorithms tending to tap too fast rather than too slow). Because of the approximate nature of ground-truth annotations, windows of precision must be used. As Figure 2.9 shows, there does not seem to be significant difference between the 4% precision used our evaluation and wider windows, however, 4% is probably the highest precision level that should be considered as the Just-Noticeable Difference (JND) for tempo differences is approximately 4% for music (Friberg and Sundberg, 1995). Further, larger precision windows may be required to evaluate *consecutive* IBIs rather than global tempo, especially in varying tempo situations (Klapuri et al., 2005). An important conclusion of this evaluation is that implementing a robust tempo induction algorithm calls for the computation of low-level frame features rather than that of onset lists as the first processing block. This can be clearly seen in Figure 2.8, which shows that algorithms based on onset detection clearly suffer more from distortion of the signal than other algorithms. However, whether this superiority of frame-based features over onset lists has any perceptual validity, as proposed by Scheirer (1998), remains to be investigated.

2.4.1.1   Open issues

Emulating the perception of the tactus by humans is still an unsolved problem. Inducing the tactus from arbitrary audio signals, without accepting alternative metrical levels, is not a solved issue, and many aspects call for further research, as listed below.

Low-level features

Which low-level features?   A common aspect of all computational models is the handling of feature lists, either as a starting point (for scores or MIDI data) or as a mid-level representation (for models that process audio). These features (e.g.

onsets, amplitudes, pitches, percussive instrument classes, frame subband energy) are assumed to convey the predominant information relevant to a rhythmic analysis. Except in the case of frame-based features, the features have a high level of abstraction, entailing an "implicit symbolism" (Scheirer, 2000). The first stages of human rhythm perception achieve a comparable parsing of auditory streams into feature lists, however, the actual modeling of these perceptual processes (the definition of perceptually relevant features) is still ongoing research.

Today's best and most robust models processing audio focus on energy in different frequency bands computed on successive short signal frames (Klapuri et al., 2005). However, many low-level features other than energy could be computed on signal frames and the determination of the audio features that serve best the task of pulse induction and further stages of rhythm description is still ongoing work. This is the object of Chapter 3.

Which frequency decomposition? Focusing on energy in different frequency bands, one might wonder how to define the frequency filterbank. Scheirer argues that his algorithm "is not particularly sensitive to the particular bands" (Scheirer, 1998). That is, the important point would be to proceed to a frequency decomposition, and not the particular choice of decomposition.

However, let us consider the algorithms that compute periodicities in frequency subbands (DixonACF, Klapuri, Scheirer and TzanetakisMM). They all use energy (or integrated amplitude) features. Of course, the performance of each system depends on the overall system, so it is hard to say anything conclusive about the best frequency decomposition (as indeed about any of the submodules). However, the fact that these systems show non-negligible differences in performance suggests that the definition of the frequency filterbank could be a significant issue, contrary to Scheirer's observation. This is investigated on page 124.

Frame values vs differential values Some pulse induction algorithms focus on energy values (e.g. TzanetakisMM) while others focus on changes from one frame to the next (e.g. estimating the derivative of frame energy values, e.g. Klapuri, or of the downsampled amplitude envelope, e.g. Scheirer). If, as Klapuri et al. (2005) claim, we assume that the difference between the use of the autocorrelation and that of comb filterbanks for pulse induction is not crucial in the performance of a tempo induction system, the performance of Scheirer vs that of TzanetakisMM[27] seems to indicate that changes in energy values would be more valuable rhythmic features than the energy values themselves. However, here also, a solid conclusion would require implementations to differ solely in this aspect. This issue is investigated further on page 126.

Periodicity functions

Which periodicity function? Based on comparisons between comb filterbanks, autocorrelation and phase-locking resonators, Klapuri et al. (2005) suggest that finding the correct periodicity function is not a key issue in meter analysis. However, there are significant differences in the accuracies obtained by AlonsoSP and AlonsoACF, which differ solely in the periodicity function block. The spectral product outperforms the autocorrelation on all data sets and all accuracy measures.[28] This finding should be verified on other data sets as the results of Alonso et al. (2004, Tables 2 and 3, p.162) seem to indicate different conclusions (namely that the autocorrelation would be better than the spectral product). A comparison with a comb filterbank method (used by the ISMIR 2004 contest winner) would also be interesting. See Chapter 4.

Combining and parsing multiple information sources

Periodicity detection before or after the integration of multiple features? Current literature (Scheirer, 1998; Klapuri et al., 2005) advocates the use of multiband processing and subsequent integration of periodicity estimates, rather than periodicity estimation after the integration of a signal processed in several frequency bands.

---

[27]Respectively 37% vs 30% with accuracy 1 and 68% vs 50% with accuracy 2

[28]Note that it is however more sensitive to distortion

The difference between TzanetakisMS and TzanetakisMM lies precisely in the integration of several frequency bands respectively before or after periodicity estimation. The algorithms exhibit similar performance when assessed with accuracy 1, but the former performs around 5% better than the latter with respect to accuracy 2. It is difficult to make any solid conclusion and confirm or refute Scheirer's point from these results. Let us however outline a few aspects of these methods: Estimating periodicities after feature integration enhances periodicities that are present in *all* features, while periodicity estimation before feature integration favors signals whose periodicities appear solely in some features (e.g. a restricted frequency band). Also, the former method has a bias towards fast metrical levels; indeed, it accounts for the phase of periodicities while the latter does not. Consider for example the case where two features have the same periodicity but have a phase difference of half the period: the former method yields double the tempo of the latter. This is verified on the data used here: TzanetakisMS makes more double-tempo errors than TzanetakisMM.[29] One can argue that each method is more suitable for different types of data. Further evaluations are required before more general conclusions can be drawn. See Chapter 4.

Joint estimation of several metrical levels Some pulse induction methods encode (implicitly or explicitly) aspects of the metrical hierarchy by letting large time-scale phenomena influence responses at smaller time scales (and inversely), e.g. comb filters. In fact, this encodes the assumption that the perception of high metrical levels, e.g. the measures, orients the perception of lower metrical levels from which they are derived. Parncutt (1994, p.434) questions this assumption, writing "each pair of events in a rhythmic sequence initially contributes to the salience of a *single* pulse sensation" (emphasis ours), and later that "pulse sensations can enhance the salience of other, consonant pulse sensations." One may understand the "initially" above as an indication not to implement influential schemes between metrical levels in the induction process, but indeed to do it in the tracking process, which is also in agreement with the Dynamic Attending Theory (Drake et al., 2000a).

---

[29]Note that accuracy 2 does not consider them as errors

Among the algorithms tested in Section 2.3, three (Klapuri, Uhle and DixonACF) implement, in different ways, explicit influential schemes for the determination of 2 or 3 metrical levels. As they all perform very well, it seems interesting to evaluate more methodically the effect of this feature. This is one of the aspects of tempo induction algorithms that Chapter 4 addresses.

Moderate tempo tendency   Similarly, the relevance of the "moderate tempo tendency" that has to be considered when focusing simultaneously on several levels, and often modeled with a prior tempo probability function, as in (Parncutt, 1994), should also be the object of further research.

Induction versus tracking   It is sometimes hypothesized that in order to compute a tempo value that best reflects human perception of the musical pace, it would be better to consider the whole tracking process rather than rely solely on tempo induction (Gouyon et al., 2004a). Performance differences between DixonT and DixonI are not really conclusive in that respect. On this point also, more research is needed.

Redundant approach to tempo induction   Section 2.3 shows several algorithms performing the same task and exhibiting specific performances on specific parts of the data. This raises an important question: Can we improve tempo estimation accuracy by combining the outputs of several algorithms? Insights into a way to address this problem are given in Chapter 6.

Better benchmarks?   Much effort is still needed to design better public benchmarks for tempo induction algorithms. Chapter 6 proposes different ways to tackle this problem.

2.4.2   Other stages of automatic rhythm description

A number of diverse formalisms have been used to implement pulse tracking models. An important aspect is the balance between inertia and reactiveness of the model.

Models with a sufficient degree of inertia can be built by accounting for several concurrent hypotheses. This seems a must for keeping the possibility of recovering after an error (preventing "garden-path errors" (Rosenthal, 1992, p.11)). Another important aspect lies in the consideration of incoming data on an event by event basis or a predicted beat by predicted beat basis. Following the former strategy is in fact making a first step towards quantizing the data, not solely tracking a pulse. AI formalisms have been proposed recently to enhance beat trackers and address quantization and pulse tracking jointly.

Very few algorithms for time signature determination exist. They usually entail the computation and parsing of a periodicity function, as in pulse induction. Apart from swing estimation, systematic timing deviation estimation is the object of few computational models. The usual rationale behind swing estimation is to consider that the tempo is constant (i.e. no long term timing deviations) and to seek predefined patterns of short term timing deviations within a pulse induction process.

Current research in rhythm description addresses all of these aspects, with varying degrees of success. For instance, recent pulse tracking systems (Dixon, 2001a; Cemgil and Kappen, 2001) reach high levels of accuracy. On the other hand, accurate quantization, score transcription, determination of time signature and characterization of intentional timing deviations are still open questions. Particularly, it remains to be seen how well recently proposed models generalize to different musical styles.

Music content processing and music information retrieval applications call for rhythmic descriptors that would entail a high level of abstraction. More research should be dedicated to this issue.

Chapter 3

Feature selection for rhythm periodicity function

In this chapter, we address one of the current research topics highlighted in Chapter 2: the determination of the low-level features of musical audio signal that convey best the rhythmic aspects of musical signals. Central to this work is the assumption that such features can be identified with the low-level features that are the most adequate for the computational identification of beat positions. In Section 3.1, we present the problem in more details and advocate the use of machine learning methods. In Section 3.2, we detail the data sets (1360 musical pieces) used in this chapter's experiments. In Section 3.3, we provide implementation details of the 274 low-level features considered. Experiments and results are reported in Section 3.4. Finally, Section 3.5 summarizes the main findings of this chapter.

3.1   Purpose and method

The purpose of this chapter is to determine which low-level features of musical audio signals are the most adequate for the computational identification of beat positions. That is, we aim at selecting among several features computed at a regular sampling rate, those whose temporal behavior would best indicate the presence and localization of beats, based on evidences observed on audio data whose beats have been previously annotated manually.

Therefore, we set up a large set of musical pieces (1360 musical pieces), whose beats have been annotated manually (there is a total of 90643 beats). In between beats, we define "non-beat" instances as detailed on page 92 —in the following, the term "instance" refers to beat as well as non-beat instances, while the term "piece" refers to a musical piece, each piece therefore contains several instances. Instances can be described by many different low-level signal features, we focus in this chapter on 274 features. Individual features and feature subsets are evaluated and ranked according to the following criteria: relevant features are those whose values are *most similar on beats* and *most dissimilar between beats and non-beats*, while irrelevant features are those who do not satisfy this property (for instance, features whose values on beats are randomly distributed, or those whose values on beats and on non-beats are similar), see Figure 3.1 for an illustration. That is, considering the two concepts, or classes, "beat" vs "non-beat," we seek features that have small within-class variation and great between-class variation. This is done according to machine learning methods.

As detailed in Section 2.2, the majority of computational systems that describe one of the many rhythmical aspect of music compute periodicities of feature lists (e.g. for pulse induction, beat tracking, etc.). This is illustrated in Chapter 5. We make the assumption that features that lead to a good separability of beat and non-beat classes will also be promising features for computing useful periodicity functions.

Let us stress here that, contrarily to Seppänen (2001), we do not view the classification experiments presented in this chapter as an actual method for finding beats in new, unknown audio signals. In our view, these experiments solely aim at providing useful information for actual beat induction and tracking algorithms, namely which low-level features to focus on (see Chapter 4 for the application of the selected features in beat induction experiments).

There exist several ways to apply machine learning techniques to the selection of representative beat features. For instance, Seppänen (2001) considers all instances at once and evaluate features on the whole data set, this amounts to seeking the best beat and non-beat models. Here, we advocate a different methodology: feature evaluation is rather done at the scope of *individual musical pieces*. Decisions are taken

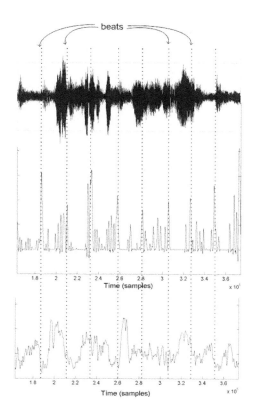

Figure 3.1: Illustration of an audio signal, its annotated beats and the temporal evolution of two features, relatively good (top, differential of the energy in the frequency band [100-216 Hz]) and bad (bottom, zero-crossing rate).

on each individual piece regarding the relevance of given features or feature subsets and then, results are integrated (averaged) over either the whole set of musical pieces or the pieces of a specific musical genre to make a final decision. A comparison of both approaches is given on page 137.

Section 3.1.1 details the method for computing beat and non-beat features from low-level features and beat annotations. Section 3.1.2 explains thoroughly the machine learning methods used in the evaluation of features and feature subsets. Section 3.2 details the data set and annotations. Section 3.3 enumerates the low-level features. Finally, Sections 3.4 and 3.5 propose results and conclusions regarding diverse feature selection experiments.

3.1.1   Computation of beat and non-beat features

Series of frame feature values can be computed at a regular sampling rate from musical pieces. Given the pieces' beat annotations, we wish to process frame features in order to derive beat features and non-beat features. The simplest way to compute beat features would be to select feature values on the closest frame of each annotated beat. A reason not to do it is that we cannot assume that beat annotations are accurate at the fine time precision of the frame rate (e.g. 10 ms). An improvement over the previous method would be to define regions of signal containing several frames around each beat and compute feature averages over these regions. However, because their purpose is to make sure to retain at least one relevant frame, such regions do contain both relevant and irrelevant frames. And there is no evidence that the mean (or any other statistics) of such frames would be relevant. We believe that it is more relevant to select a single frame in such regions (the "beat frame") and propose to focus on the frame with the *maximal* value.

On the other hand, non-beat features are defined on frames chosen randomly between each pair of beats. The computation of beat and non-beat features is illustrated in Figure 3.2.

Of course, different ways to compute beat and non-beat features could be devised, and assumptions underlying our method could be discussed. For instance, is it

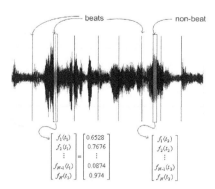

Figure 3.2: Example of audio signal and imaginary beat and non-beat features.

reasonable to consider that beat features can be measured on a single signal frame?
Should the computation of beat feature focus on frames with the maximal value in
each beat region of a specific piece rather than on frames with the *most similar* values
among beat regions?[1] This is left for future work.

3.1.2 Feature selection methods

In pattern recognition research, feature selection have become the focus of great
interest. In many recent application domains, researchers must now deal with datasets
described by huge numbers of features (e.g from hundreds to tenth of thousands
of features in text processing, image processing, bioinformatics, genomic analysis).
Feature selection differs from feature transformation (as e.g. Principal Component
Analysis). In the latter, new features are created by transforming the original set of
features (by linear or non-linear combinations) while in the former some features are
kept and some others discarded. As Guyon and Elisseeff (2003) illustrate, there are
several reasons for reducing the number of features: getting simpler models (faster
and less computationally-expensive classifiers), improving prediction accuracy (i.e.

---

[1]Each piece having several beat regions and each such region having several candidate frames,
the best path (in terms of similarity) could be found by applying the Viterbi algorithm.

defying the "curse of dimensionality" (Jain et al., 2000)) and providing more insights into which aspects are relevant (better understanding of the processes that generated the data).

Recall that, in our context, the main purpose is not the building of a classifier; it is rather the selection of the features that yield the best rhythmic periodicity functions. We will not finally build a "beat model," i.e. a classifier that would predict the membership of some unknown data to either the "beat" class or the "non-beat" class. As already mentioned, the main assumption of this work is that the best features in the classification experiments described in this chapter will also be the most promising features for computing a periodicity function suitable for pulse induction, pulse tracking and other rhythmic applications. Therefore, here, feature selection is mostly interesting for reasons of data understanding and computational cost.

In some cases, feature selection can be driven by our understanding of the feature meanings and our intuitions regarding their relevances. In addition, automatic attribute selection methods can be used, reviews of which can be found in (Dash and Liu, 1997; Blum and Langley, 1997; Kohavi and John, 1997; Guyon and Elisseeff, 2003; Liu and Yu, 2005). Such methods account for two fundamental processes: *feature subset generation* and *subset evaluation*. The most straightforward approach to subset generation would be to consider the exhaustive list of the possible combinations of features into subsets. However, as the number of possible subsets grows exponentially with the number of features, this is impractical for even moderate numbers of features. Therefore, subset generation is essentially a heuristic search problem, with each state in the search space specifying a candidate subset (Blum and Langley, 1997; Kohavi and John, 1997). In the exploration of the state space, two problems must be addressed: the search *direction* (growing subset vs shrinking subsets) and search *strategy*: complete search (e.g. best-first heuristics), sequential search (consider one feature at a time, e.g. "greedy" search) or random search (e.g. genetic search) (Liu and Yu, 2005). The search strategy has an impact on the type of output, which can be either a ranked list of features or a bag of unranked features.

Given a subset generation procedure, different evaluation methods can be used.

Subset evaluation methods can be classified in two families: *filters* vs *wrappers* (Kohavi and John, 1997). In the filtering approach, the relevance of a feature subsets is evaluated from general characteristics of the data, without implying any classifier. According to Kohavi and John (1997) and Blum and Langley (1997) there exist various notions of feature relevance. For instance, features can be evaluated with respect to correlation with the classes and inter-correlation between features. Other relevance measures include e.g. separability measure and information gain (Hall and Holmes, 2003). In the wrapping approach, feature subsets are evaluated according to the prediction accuracy of a given classifier used as a black box.

In addition to the aspects of subset generation and evaluation, Liu and Yu (2005, p. 494) also propose to categorize automatic feature selection algorithms with respect to a third aspect, the *learning context*: supervised vs non-supervised. The former denotes tasks for which class membership is known for all instances (e.g. classification) while the latter corresponds to tasks for which class membership is unknown (e.g. clustering). We will not focus further on this distinction as our context is restricted to classification.

In the following experiments, feature and feature subset *evaluation* follow the *wrapper* approach to feature selection. We use different classifiers (corresponding to different learning strategies: instance-based learning, statistical modeling or decision tree building) which are presented in more details on the following page.

In most cases, subset *generation* is manual. There are several reasons for that: we have some domain knowledge, features can be naturally grouped by "family," hence the will to evaluate the goodness of each feature family; and it seems relevant to evaluate and compare features promoted in the current literature. Some experiments also imply automatic subset generation, the method used is described on page 99.

Most classification accuracies reported in this chapter are computed as 10-fold cross-validations (or indicated otherwise): a subset containing 90% randomly selected samples from the data set is selected for learning and the remaining 10% are used for testing, this is repeated 10 times, the final accuracy is the average over those 10 runs (see illustration on Figure 3.3). Note that in our case, 10-fold cross-validations are computed on *individual* pieces. Each piece contains between 20 to around 300 instances (see the distribution of the number of beats per piece on Figure 3.5). An accuracy estimation of a given feature subset is obtained for each piece, the final accuracy estimation is then computed as the average over the whole set of pieces (or the pieces of a specific musical genre, when indicated).

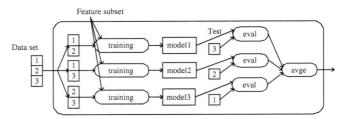

Figure 3.3: Illustration of the cross-validation method for accuracy estimation (3-fold cross-validation shown, adapted from (Kohavi and John, 1997, Figure 6)).

3.1.2.1 Classifiers

Experiments described in this chapter have been conducted with the free software Weka (Witten and Frank, 2000), available under GPL license at http://www.cs.waikato.ac.nz/ml

An important distinction has to be made between induction algorithms that consider all features as equally important (as e.g. Naive Bayes) and those that either put different weights to features or incorporate embedded feature selection schemes (as e.g. decision trees). The former family gives an estimate of the goodness of selected feature subsets. The latter rather gives an estimate of the goodness of some features inside a selected subset. While the former seems more relevant to the evaluation of a subset as a whole, the latter can be relevant in case the subset will be subsequently postprocessed.

C4.5 Decision trees (see an illustration in Figure 3.4) are hierarchical decision structures made up of branches, nodes and leafs (Witten and Frank, 2000, p.58). A node

---

[2]We used Weka version 3.4.4.

involves testing a particular attribute (as for instance the humidity in Figure 3.4, which can take the values "high" or "normal"). Leaf nodes give a classification for all instances that reach the leaf. For example, any instance whose attribute values are "Outlook=sunny" and "Humidity=high" is classified as a member of Class1. In the case of numeric attributes (which concerns us here), the test at the node determines whether the value is greater or less than a given constant (determined during the training phase).

See (Witten and Frank, 2000, p.89) for details on the problem of constructing a decision tree from a set of training examples. Note however that decision trees usually put different weights on attributes or incorporate embedded attribute selection schemes (as does e.g. C4.5). Note also that decision trees are usually sensitive to irrelevant features (Kohavi and John, 1997).

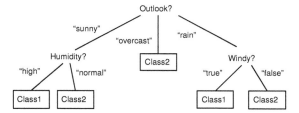

Figure 3.4: Example of an imaginary decision tree (from (Witten and Frank, 2000, p.92)), with attributes "Outlook?", "Humidity?" and "Windy?" and two possible classes, which may correspond to playing sports or no (respectively Class2 and Class1).

In the experiments detailed below, we used the algorithm j48, which is the Weka implementation of C4.5 version 8 (Quinlan, 1993), with the default parameter settings (importantly, a confidence factor of 0.25 and a minimum number of instances per leaf of 2). The minimum number of instances is relatively small and one may object that it may lead to overfitting. However, as our purpose here is not to seek good generalization of the beat models,[3] this is not an issue. Attributes are normalized

---

[3]but rather good description of individual pieces, the subsequent process of averaging over many pieces guarantees that features that are good only on very particular cases are ruled out

beforehand.

DecisionStump   A "decision stump" is a simple decision tree that have only one level, that is, which makes decision with respect to a single feature (Witten and Frank, 2000, p.257).

k-NN   k-Nearest Neighbors is an example of instance-based learning. The basic idea in instance-based learning is to memorize a set of training instances, the majority class of the k training instances that most strongly resembles a new incoming instance is assigned to this new instance (Witten and Frank, 2000, p.72). Unlike algorithms that incorporate embedded feature selection techniques, as decision trees, k-NN accounts by default for all attributes in the computation of the distance between two instances. This causes this classification scheme to be sensitive to irrelevant features (Kohavi and John, 1997).

Here, the definition of the distance is the standard Euclidean distance. We used Weka's implementation with the following parameter settings: the attributes were normalized, the number of neighbors was set to 3 (because of the noisy nature of the data) and no distance weighting was used.

NaiveBayes   NaiveBayes classifiers are based on the the Bayes rule, which states that given a class ω_i and a set of attributes values x,

$$p(\omega_i \mid x) = \frac{p(x \mid \omega_i)p(\omega_i)}{p(x)}$$

The notation $p(A)$ denotes the probability of A and $p(A \mid B)$ denotes the probability of A *given* B. Bayes rule can be also expressed informally as *posterior* $\propto$ *likelihood* $\times$ *prior*. That is, the probability of class ω_i given an instance attribute values is proportional to the probability of these values when we know the instance pertains to class ω_i, multiplied by the posterior probability of class ω_i. Assuming that all attributes are independent, it is possible to write that

$$p(x \mid \omega_i) = \prod_{k=1}^{N} p(x_k \mid \omega_i)$$

where N is the number of attributes. Given a specific value for an attribute, its likelihood $p(x_k \mid \omega_i)$ given class ω_i is observed has to be estimated. The usual way to do that is to assume that attributes follow Gaussian probability distributions, whose means and variances can be estimated on training data.

Note that the NaiveBayes classification scheme computes class membership by considering all attributes as equally important (they all have the same importance in the numerator multiplication). This, together with the strong (and usually unrealistic) attribute independence assumption, are the reason for the name "Naive" Bayes. However, this method has shown to work very well in many real situations. See (Witten and Frank, 2000, pp.82-89) and (Duda et al., 2001, Chap.2) for more details on this method.

Note that unlike decision trees and instance-based classifiers, Naive Bayes classifiers are robust to irrelevant features (but not to correlated features (Kohavi and John, 1997)).

We used Weka's implementation with default parameter settings and normalized attributes beforehand.

3.1.2.2 Automatic subset generation

As already mentioned, feature subset generation is manual in most experiments. A few experiments however used an automatic subset generation method (see on page 128 and 131). We used the *forward best-first search* procedure implemented in Weka (with default parameter settings), that is, growing subset are considered. Forward search can be justified when the ratio between relevant features and the total number of features is assumed to be small (Liu and Yu, 2005, p.497).

In these experiments, subset evaluation was done by wrapping with either k-NN or NaiveBayes (both gave similar results). We used Weka's algorithm "WrapperSubsetEval." This algorithm evaluate the worth of a subset by five-fold cross-validations (see Figure 3.3) that are repeated multiple times. The number of repetitions is determined by looking at the standard deviation of the accuracy estimate: if it is higher than 0.01, another cross-validation run is executed (see (Kohavi and John, 1997) and Weka source code documentation).

3.2 Data and associated metadata

In this section, we describe the musical pieces used for the beat feature selection experiments detailed in this chapter.

The data used in this chapter comes from different sources (personal collections and publicly-available data) and comes with different types of legacy metadata.[4]

There is a total of 1360 audio files which together amount 90643 beats (with a minimum of 7 beats per piece and a maximum of 262 beats per piece, see distribution of beats per piece on Figure 3.5). 89283 non-beats have been defined as detailed on page 92.

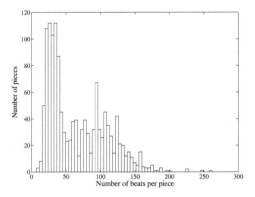

Figure 3.5: Distribution of the number of beats per piece, feature selection dataset.

[4]Here again, we wish to thank those who helped in the recollection of this data and metadata: Anssi Klapuri, Matthew Davies, Stephen Hainsworth and Giorgos Emmanouil.

CUIDADO excerpts This set is made up of 70 audio files, in .wav format, ripped from commercial CDs. Files are around 20 s each. Some have varying tempo. This data set has been collected and annotated during the CUIDADO project[5] at the MTG, it has been used by Gouyon and Herrera (2003b) for Meter induction experiments. It comes with beat annotations at two metrical levels: the tactus and the measure, only the former has been used here. Annotations have been put by a single person and have not been cross-checked. Audio data is not publicly available for copyright reasons. The genre distribution is detailed in Table 3.1.

Genre	# pieces
Acoustic	24
Classical	13
Jazz Fusion	4
Rock/Pop	29

Table 3.1: Genre distribution of the CUIDADO excerpts

SIMAC excerpts This set is made up of 595 audio files, in .wav format, ripped from commercial CDs. Files are around around 20 s each. Some files (a minority) have varying tempo. This data have been collected and annotated during the SIMAC project (http://www.semanticaudio.org) at the MTG and was used internally by MTG researchers in different audio description experiments. Among other annotations, this data comes with tempo and beat positions. Beats have been positioned by a professional musician and cross-checked by the author of this book. The ground-truth tempo was computed as the median of the IBIs. 465 of these files were used for the tempo induction contest reported in (Gouyon et al., 2006) (see also page 66). The remaining 130 files were added after the contest. Part of this data is available on the Internet (see page 66). The genre distribution is detailed in Table 3.2:

Hainsworth's excerpts This set is made up of 221 audio files, in .wav format, ripped from commercial CDs. File length are between 11 s and 1 mn 30 s. Some have

---

[5]http://www.upf.edu/ec/proyectoseuropeos/cuidado2.htm

Genre	# pieces
AfroBeat	3
Balkan and Greek	144
Classical	84
ClassicalSolo	72
Electronic	59
Fado	10
Flamenco	13
Jazz	56
Latin	44
Rock	68
Samba	42

Table 3.2: Genre distribution of the SIMAC excerpts

varying tempo. This data set has been used by Hainsworth (2004) in beat tracking experiments. It comes with beat annotations at one metrical level and genre annotations in 11 classes. The genre distribution is detailed in Table 3.3.[6] Annotations have not been cross-checked. Audio data is not publicly available for copyright reasons.

Genre	# pieces
Acoustic Folk	4
Classical	23
ClassicalSolo	7
Choral	21
Dance	40
Folk	18
Jazz	31
Big band Jazz	9
Rock/Pop	30
Pop 60's	38

Table 3.3: Genre distribution of Hainsworth's excerpts

---

[6] Acoustic Folk was originally "random stuff" and "unused"

Klapuri's excerpts This set is made up of 474 audio files, in `.wav` format, ripped from commercial CDs. File length are between 23 s and 1 mn 56 s. Some have varying tempo. This data set has been used by Klapuri et al. (2005) in beat tracking experiments. It comes with beat annotations at one metrical level and genre annotations in 7 classes. The genre distribution is detailed in Table 3.4. Annotations have not been cross-checked. Audio data is not publicly available for copyright reasons. Further data statistics are available at `http://www.cs.tut.fi/~klap/iiro/meter/database.html`.

Genre	# pieces
Acoustic Folk	15
Classical	84
Electronic/Dance	66
Hip-Hop/Rap	37
Jazz/Blues	94
Rock/Pop	122
Soul/RnB/Funk	56

Table 3.4: Genre distribution of Klapuri's excerpts

3.2.1 Musical genres

Original genres were grouped in 10 genres, mostly with respect to their instrumental or rhythmic contents:

- *Acoustic*: Klapuri's and Hainsworth's *Acoustic Folk*, Hainsworth's *Folk*, CUIDADO *Acoustic* and SIMAC *Fado* and *Flamenco* excerpts (84 pieces in total, mostly sung pieces accompanied with acoustic instrument, as the guitar, presence of some percussion but no loud drums)

- *Jazz/Blues*: Klapuri's *Jazz/Blues*, Hainsworth's *Jazz* and *Big band Jazz*, CUIDADO *Jazz Fusion* and SIMAC *Jazz* excerpts (194 pieces in total, quite heterogeneous set, many instrumental pieces with lots of horns, many jazz-like drum playing style)

- *Classical*: Hainsworth's, Klapuri's, CUIDADO and SIMAC *Classical* excerpts (204 pieces in total, mostly orchestral music, symphonies or sonatas, few operas)

- *Classical solo*: Hainsworth's and SIMAC *Classical solo* excerpts (79 pieces in total, pieces for solo instruments as piano, guitar or organ)

- *Choral*: solely from Hainsworth (21 pieces, just choirs)

- *Electronic*: Hainsworth's *Dance*, Klapuri's *Electronic/Dance* and SIMAC *Electronic* (165 pieces, lots of electronic drums, mostly strong beats)

- *Afro-American*: Klapuri's *Hip Hop/Rap* and *Soul/RnB/Funk* (93 pieces, $\frac{4}{4}$ time signatures and clear drum patterns on the great majority of excerpts: low-frequency bass drum on beats 1 and 3 and brighter snare drum on beats 2 and 4)

- *Rock/Pop*: Hainsworth's *Rock/Pop* and *Pop60s*, Klapuri's and CUIDADO *Rock/Pop* and SIMAC *Rock*, *Latin* and *AfroBeat* (334 pieces, quite heterogeneous set, mostly sung pieces, with drums)

- *Balkan/Greek*: solely from SIMAC (144 pieces, sung pieces accompanied by acoustic instruments, typical folklore music from Greece and Balkans)

- *Samba*: solely from SIMAC (42 pieces, sung pieces accompanied by acoustic instruments, with typical second and fourth beats marked by a low-frequency percussion, typical folklore music from Brasil)

3.3 List of features

For all the features, the frame size is set to around 23.2 ms and the hop to around 11.6 ms (respectively 1024 and 512 samples at a sampling frequency fs of 44100 Hz), that is, the overlap is 50%. Feature sampling rate is therefore $44100/512 = 86.1$ Hz.

In this Section, a total of *274 features* are described.

3.3.1 Spectral and Cepstral features

On each frame, a total of *33 features* are computed with the free software CLAM.[7] They are described below.[8] They will be referred to as "CLAM features."

Audio features From the samples of waveform signal chunks, the following feature is computed:

Zero-Crossing Rate (ZCR) The ZCR is a measure of the number of waveform time-domain zero-crossings (i.e. sign changes), averaged over the number of samples.

Spectral features These features are computed in the spectral domain. Frame samples are multiplied with a Hamming window and a discrete Fourier transform is subsequently computed by means of the FFT. All the following features are computed on the magnitude spectrum X corresponding to positive frequencies, elements of X are denoted X_i below, with $i = 1...N$ (N is half the number of samples in a frame, in our settings, $N = 512$).

Spectrum mean Considering the magnitude spectrum X_i as a distribution, the arithmetic mean is computed as

$$mean(X) = \frac{\sum_{i=1}^{N} X_i}{N}$$

Spectrum spread The definition of the spectrum spread used here is the variance of the spectrum around its mean value.

$$spread(X) = \frac{\sum_{i=1}^{N} (X_i - mean(X))^2}{N}$$

---

[7]http://www.iua.upf.es/mtg/clam
[8]Thanks to David Garcia for writing a first version of this subsection on CLAM features.

Spectrum geometric mean   The geometric mean is the $N_t h$ root of the product of N elements.

$$gMean(X) = \left(\prod_{i=1}^{N} X_i \right)^{1/N}$$

It can also be computed as[9]

$$\log(gMean(X)) = \frac{\sum_{i=1}^{N} \log X_i}{N}$$

Spectrum energy   The energy is computed as

$$e(X) = \sum_{i=1}^{N} X_i^2$$

Spectrum centroid   It can be understood as the center of gravity, or barycenter, or again "equilibrium point" of the spectrum magnitude distribution, it is computed as

$$centroid(X) = \frac{\sum_{i=1}^{N} i X_i}{\sum_{i=1}^{N} X_i}$$

Spectrum flatness   The spectral flatness is the ratio between the geometric mean and the mean. It gives a measure of the spectrum flatness along frequency.

$$flatness(X) = \frac{gMean(X)}{mean(X)}$$

Spectrum magnitude kurtosis   It is based on the spectrum *magnitude* distribution $p(X)$ (not the spectrum distribution X, see Figure 3.6), elements of $p(X)$ are denoted $p_k(X)$ where $k = 1...N_{p(X)}$, $N_{p(X)}$ being the number of elements of $p(X)$.

It is a measure of the magnitude distribution flatness, or respectively peakedness, around its mean value. Distributions whose tails on both sides on the mean are wider than the Normal distribution (i.e. flat-topped distributions) have a kurtosis smaller than 3. Those with smaller tails (i.e. peaky distributions) have a kurtosis greater

---

[9]where log can be the logarithm in base e or 10

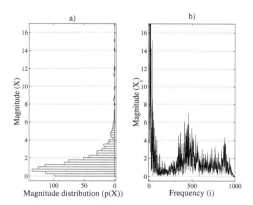

Figure 3.6: Example of magnitude spectrum (b) and spectrum magnitude distribution (a).

than 3.

$$kurtosis(p(X)) = \frac{\sum_{k=1}^{N_{p(X)}} \left((p_k(X) - mean(p(X)))^4 \right)}{\left(\sum_{k=1}^{N} (p_k(X) - mean(p(X)))^2 \right)^2}$$

Spectrum magnitude skewness It is a measure of the degree of asymmetry of the spectrum magnitude distribution around its mean. If the distribution tail in the positive direction is longer than that in the negative direction, the skewness is positive, in the opposite case the skewness is negative. The Normal distribution has 0 skewness.

$$skewness(p(X)) = \frac{\sum_{k=1}^{N_{p(X)}} \left((p_k(X) - mean(p(X)))^3 \right)}{\left(\sum_{k=1}^{N_{p(X)}} (p_k(X) - mean(p(X)))^2 \right)^{\frac{3}{2}}}$$

Spectrum maximum magnitude frequency This is the frequency of the highest magnitude in the spectrum

Spectrum low-frequency energy relation This is the ratio of the spectrum magnitude energy below 100 Hz to the total energy.

Spectrum rolloff This is the frequency below which 85% of the signal energy remains. It is denoted f_r in the following equation:

$$\sum_{i=1}^{f_r \frac{N}{f_s/2}} X_i^2 = 0.85 \sum_{i=1}^{N} X_i^2$$

Spectrum high frequency content (HFC) This is comparable to the total energy, at the difference that energies at high frequencies are given more importance than magnitudes at low frequencies. The weighting is linear along the frequency axis.

$$HFC = \sum_{i=1}^{N} i X_i^2$$

Spectrum slope This is the slope of the linear regression that best fits the magnitude spectrum. A negative slope indicates globally decreasing spectral magnitudes.

Mel-Frequency Cepstral Coefficients (MFCCs) MFCCs are widespread descriptors in speech research. The Cepstral representation has been shown to be of prime importance in this field, partly because of its ability to nicely separate the representation of voice excitation (the higher coefficients) from the subsequent filtering performed by the vocal tract (the lower coefficients). Roughly, lower coefficients represent spectral envelope (i.e. the formants) while higher ones represent finer details of the spectrum, among which the pitch (Oppenheim and Schafer, 2004). One way of computing the Mel-Frequency Cepstrum of a spectrum is as follows:

1. Projection of the frequency axis from linear scale to the Mel scale, of lower dimensionality (i.e. 20, by summing frequency-bin powers within each triangularly-weighted band of a Mel critical band filterbank)

2. Computation of the logarithm of Mel-band power values

3. Discrete Cosine Transform (DCT)

The number of output coefficients of the DCT is variable. Here, it is set to 13, as in the standard implementation of the MFCCs detailed in the widely-used speech processing software Hidden Markov Model Toolkit (HTK, version 3.2.1).[10]

Spectral peak features The magnitude spectrum is further processed by a peak-picking algorithm and a partial track estimation algorithm (hence accounting for frame-to-frame context) that parse local maxima of the spectrum into harmonic peaks (corresponding to harmonics of an instrument) and spurious peaks (due e.g. to noise). The algorithm implementation available in CLAM is based on the technique described in (Serra, 1989, pp.42-48). Other descriptors are computed on the series of spectral peak amplitudes corresponding to each frame (noted a_i below, with N_P being the number of peaks.

Spectral peak magnitude mean This is the arithmetic mean of the peak magnitudes. The formula is similar to that of the spectrum mean above.

Harmonic centroid This is the center of gravity of the peak magnitudes. The formula is similar to that of the spectrum centroid above.

Harmonic deviation This is the sum of the absolute deviation of peaks with respect to local spectrum envelopes, normalized by the sum of the peak magnitudes. Local spectrum envelopes (SE_i below) are computed as the mean of the magnitudes of 3 consecutive peaks. It is computed as

$$deviation = \frac{\sum_{i=1}^{N_P} |a_i - SE_i|}{\sum_{i=1}^{N_P} a_i}$$

---

[10]http://htk.eng.cam.ac.uk/

First tristimulus   This is the ratio of the first peak energy to the sum of all peak energies. It is computed as

$$first\ tristimulus = \frac{a_1^2}{\sum_{i=1}^{N_P} a_i^2}$$

Second tristimulus   This is the ratio of the sum of the second, third and forth peak energies to the sum of all peak energies. It is computed as

$$second\ tristimulus = \frac{a_2^2 + a_3^2 + a_4^2}{\sum_{i=1}^{N_P} a_i^2}$$

Third tristimulus   This is the ratio of the sum of all peak energies from the fifth to the sum of all peak energies. It is computed as

$$third\ tristimulus = \frac{\sum_{i=5}^{N_P} a_i^2}{\sum_{i=1}^{N_P} a_i^2}$$

3.3.2   Onset detection functions

In this Paragraph, *5 features* are described. They are based on onset detection function Matlab code gently provided by Juan Bello and colleagues from Queen Mary University of London (QMUL). They will be referred to as "QMUL features."

As in Paragraph 3.3.1, the audio signal is chunked in frames of 1024 samples, with 50% overlap. Then frame samples are multiplied with a Hanning window (not Hamming as on page 105) and the FFT is computed. Then, features are computed on the magnitude spectrum X corresponding to positive frequencies, elements of X are denoted X_i below, with $i = 1...N$ (N is half the number of samples in a frame, in our settings, $N = 512$).

- Spectral difference (SD): The implementation is based on that of (Duxbury et al., 2002)

$$SD(X) = \sum_{i=2}^{N} \sqrt[2]{|X_i^2 - X_{i-1}^2|}$$

- High-frequency content (HFC): This is the sum of the spectrum magnitudes, with a linear weighting scheme giving more importance to high frequencies than to lower ones. Contrastingly with the HFC detailed on page 108, here, the HFC is computed from magnitudes, not energies.

$$HFC_2(X) = \sum_{i=1}^{N} iX_i$$

- Energy: At the difference with the energy computed on page 106, here, the energy is normalized and more importantly is expressed on the dB scale instead of the linear scale. It is computed as

$$e_2(X) = 20\log_{10}\left(\frac{1}{N}\sum_{i=1}^{N}X_i^2\right)$$

- Phase deviation: A measure of the shape of the distribution of phase deviations between successive frames defines another onset detection function. Unlike the detection function based on energy, which favors percussive onsets, this detection function favors soft onsets. See (Bello and Sandler, 2003) for more details.[11]

- Complex spectral difference: Energy-based and phase-based detection functions are complementary and focus on different types of onsets, Bello et al. (2004) combine them into another detection function that computes spectral difference between successive frames directly in the complex domain.

Note that three of these features account a differentiation operator, they are therefore not further processed as detailed in Section 3.3.5.

3.3.3 Subband energies

Dixon-like In this Paragraph, a total of *8 features* is described.

Dixon et al. (2003) and Paulus and Klapuri (2002) compute eight feature lists ("mid-level representation") representing the audio signal in eight non-overlapping

---

[11]Note that it is related to the "group delay" feature proposed by Sethares et al. (2005).

frequency bands. A bank of eight 6th order Butterworth filters is designed as follows: a first low-pass filter with a cut-off frequency of 100 Hz and 6 band-pass filters and 1 high-pass filter distributed uniformly on a logarithmic frequency scale (i.e. passbands are approximatively [100 Hz – 216 Hz], [216 Hz – 467 Hz], [467 Hz – 1009 Hz], [1009 Hz – 2183 Hz], [2183 Hz – 4719 Hz], [4719 Hz – 10200 Hz] and [10200 Hz – 22050 Hz]).

Then, in each band, after group delay has been taken into account,[12] the signal is half-wave rectified, squared, decimated (to a sampling frequency of 980 Hz, i.e. a decimation factor of 45 if the original sampling frequency is 44100 Hz) and smoothed with a 4th order Butterworth low-pass filter with a cutoff frequency of 20 Hz (in the downsampled domain). Here also, group delays are accounted for in each band. Finally, the dynamic range is compressed with a logarithmic function.

This process is not frame-based, however, in our understanding, it is comparable to a frame-based process: the hop size would be around 1 ms (the inverse of the sampling frequency, i.e. 980 Hz) and the frame size would depend on the size of the impulse responses of the filters applied on the original signal.

For the experiments described in this chapter, we use a slightly different version for these features, with a hop size of around 6 ms (256 samples at 44.1 kHz) and a frame size of around 12 ms (512 samples at 44.1 kHz). The computation follows the same steps as described above, but with a decimation factor of 32 (yielding a sampling frequency of around 1380 Hz if the original sampling frequency is 44100 Hz) and the dynamic range is not compressed. Then, the resulting time series is downsampled as follows: each 16 samples ($\frac{desiredHopSize}{decimationFactor} = \frac{512}{32}$), the values of 32 ($\frac{frameSize}{decimationFactor} = \frac{1024}{32}$) samples are summed.

This is comparable to a frame-based computation, in the time domain, of the *energy* of half-wave rectified subband signals.

Scheirer-like In this Paragraph, a total of *6 features* is described.

Scheirer (1998) computes feature lists also representing the audio signal in non-overlapping frequency bands. Six sixth-order elliptic filters are designed, the cutoff frequency are the following: [0 Hz – 200 Hz], [200 Hz – 400 Hz], [400 Hz – 800 Hz],

---

[12]by left-shifting the signal to an appropriate amount depending on the specific band

[800 Hz – 1600 Hz], [1600 Hz – 3200 Hz] and [3200 Hz – 22050 Hz]). They all have 3 dB of ripple in the passband and a stopband 40 dB down from the peak value in the passband. Elliptic filters show steeper transition bands than Butterworth filters, as those used on page 111, but have higher ripple in the passband.

Then, in each band, the signal is half-wave rectified and convolved with a 200 ms half-Hanning window, decimated to a sampling frequency of around 100 Hz, first-order differentiated and finally half-wave rectified to yield the subband signal amplitude envelope.

This process is not frame-based. However, in our understanding, convolving with a half-Hanning window and downsampling is similar to a framing of the signal (with a frame-size equal the window size, i.e. 200 ms) and a hop size of e.g. 10 ms if the downsampling frequency is 100 Hz.[13]

We use a slightly different implementation of these features, more similar to those described on page 111. The subband signals are computed as in the original paper. Then the decimation process is achieved first. As on page 111, the decimation factor is 32 (yielding a sampling frequency around 1380 Hz instead of 100 Hz). Then, for reasons of efficiency we use a low-pass filter instead of the convolution window (a 4th order Butterworth filter with a cutoff frequency of 10 Hz in the downsampled domain, to emulate the low-pass behavior of the 200 ms half-Hanning window), the group delays are taken into account. The signal is then half-wave rectified. Finally, the resulting time series is further downsampled as on page 111 so that the data reduction with respect to the original subband signal reaches a factor of 512, i.e. the final sampling frequency is around 86.1 Hz. In contrast to Scheirer's version, no differentiation is achieved (see Section 3.3.5 for this matter).

Apart from the frequency filterbank definition, the differences with the method described on page 111 are the following: the half-wave rectification is achieved at the end (not as the first step), the signal is not squared (the features represent subband amplitude, not energy) and the low-pass smoothing filter has a cutoff frequency of 10 Hz instead of 20 Hz.

---

[13]13.3 ms when downsampling at 75 Hz, value advocated by Scheirer (1998) for reaching real-time performances.

3.3.4 ERB-based features

In this Section, a total of *44 features* is described.

Klapuri et al. (2005) defines a set of 4 frame-based features based on the computation of the signal power in different subbands. At the difference with Section 3.3.3, the processing is done in the frequency domain. The signal is chopped into frames of 23 ms which are Hanning-windowed and overlap 50%. After zero-padding to 46 ms, the discrete Fourier transform of each frame is computed and multiplied with 36 triangular-response overlapping bandpass filters distributed between 50 Hz and 20 kHz as follows: [50 Hz – 122 Hz], [84 Hz – 166 Hz], [123 Hz – 214 Hz], [165 Hz – 269 Hz], [214 Hz – 330 Hz], [269 Hz – 398 Hz], [330 Hz – 475 Hz], [398 Hz – 562 Hz], [475 Hz – 659 Hz], [562 Hz – 768 Hz], [659 Hz – 890 Hz], [768 Hz – 1028 Hz], [890 Hz – 1182 Hz], [1028 Hz – 1355 Hz], [1182 Hz – 1549 Hz], [1355 Hz – 1768 Hz], [1549 Hz – 2013 Hz], [1768 Hz – 2288 Hz], [2013 Hz – 2597 Hz], [2288 Hz – 2944 Hz], [2597 Hz – 3333 Hz], [2944 Hz – 3771 Hz], [3333 Hz – 4262 Hz], [3771 Hz – 4813 Hz], [4262 Hz – 5432 Hz], [4813 Hz – 6126 Hz], [5432 Hz – 6907 Hz], [6126 Hz – 7782 Hz], [6907 Hz – 8766 Hz], [7782 Hz – 9870 Hz], [8766 Hz – 11110 Hz], [9879 Hz – 12501 Hz], [11110 Hz – 14064 Hz], [12501 Hz – 15819 Hz], [14064 Hz – 17788 Hz] and [15819 Hz – 20000 Hz]. Then, in each band, the signal is squared and summed.

To design the filters,[14] the linear frequency scale is mapped onto the ERB (Equivalent Rectangular Bandwidth) critical band scale by the following equation $ERB(f) = 21.2655\log_{10}(0.004368f + 1)$ (Moore, 1995, p.176, equation 10), the ERB scale is then uniformly sampled in 36 bins whose boundaries define filter center frequencies and bandwidths as illustrated in Figure 3.7.[15]

Three feature groups are then computed. First, the energy values in the 36 frequency bands (*36 features*). Second, the sum of adjacent bands by groups of 9 (yielding *4 features*). Third, the features proposed by Klapuri et al.: in each band, the degree of change in the energy values is estimated as the differential of energy normalized with energy and half-wave rectified as detailed in Section 3.3.5.[16] The degree of

---

[14]no design of filter *coefficients* takes place here

[15]Thanks to Anssi Klapuri for the subband decomposition Matlab code.

[16]As explained on page 115, implementation slightly differs from (Klapuri et al., 2005).

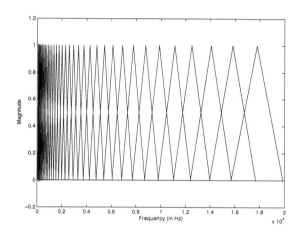

Figure 3.7: ERB frequency subbands.

change is then weighted and averaged with the magnitude (see (Klapuri et al., 2005, equation 3)).[17] Finally, adjacent bands are summed by groups of 9 to get 4 feature vectors (*4 features*).

3.3.5 Feature differentials

In this Section, a total of *178 features* is described.

Some algorithms published in the literature focus on feature values computed on frames, e.g. energy in subbands (Tzanetakis and Cook, 2002; Paulus and Klapuri, 2002; Dixon et al., 2003) while others focus on degree of change of feature values over time. This can be computed for instance as the derivative of feature values, which can be estimated by a first-order differentiator (Scheirer, 1998; Laroche, 2003) or more accurately as proposed in (Alonso et al., 2004).

The degree of change can also be measured as the feature differential normalized with its magnitude, e.g. (Klapuri et al., 2005). This is supposed to provide a better

---

[17]At the difference with (Klapuri et al., 2005), both are normalized with respect to their respective maximum values before averaging.

emulation of human audition, indeed, according to Weber's law, for humans, the just-noticeable-difference in the increment of a physical attribute depends linearly on its magnitude before incrementing. That is, $\Delta x_t / x_t$ (where x_t is a specific feature and Δx_t is the smallest perceptual increment) would be constant.

In the experiments described below, in addition to the 96 features described between page 104 and page 115, other features have been designed by computing the differential of the 89 features corresponding to instantaneous time series (i.e. leaving aside the 7 features that already correspond to differentials: the 4 features from (Klapuri et al., 2005) and 3 out of the 5 onset detection functions). This has been done in two ways (therefore yielding $89 \times 2 = 178$ features). First, as a first-order differentiation of feature values and half-wave rectification. Second, as Klapuri et al. (2005, equation 2) propose, as the half-wave rectification of the first-order differential of the μ-law compression of x_t ($y_t = \log(1 + \mu x_t)/\log(1 + \mu)$), where x_t is the time-varying feature at hand, μ is a constant (set to 100) and log is the natural logarithm (base e). Indeed, $(d/dt)[\log(x_t)] = [(d/dt)(x_t)]/x_t$, and the μ-law compression behaves like $\log(x_t)$ except around $x_t = 0$ where it permits to avoid numerical problems.[18]

3.4 Results

In order to determine relevant and irrelevant features with respect to the presence and localization of beats in musical signals, several experiments are conducted. These experiments aim at providing answers to some open questions in current literature listed in Section 2.4, they are presented in turn in the following Sections.

As we defined the same number of beats and non-beats for each piece (to be precise, there is always exactly one less non-beat than beat per musical piece), the *baseline* is around 50% for each file (this is the classification rate when always guessing the most probable class). This value should be kept in mind when assessing the goodness of any feature set. For instance, any feature subset, associated to a given classifier,

---

[18]Our approach here differs from Klapuri et al.'s in that we do not upsample nor lowpass y_t (yielding z_t in (Klapuri et al., 2005)), neither do we compute a weighted average u_t of z_t and its differential (Klapuri et al., 2005, equation 3).

yielding an accuracy value lower than 50% would be less effective than just flipping a coin (accuracy estimation is explained on page 96).

Recall that prediction accuracies are computed on each musical piece and then averaged over either the whole data set or the pieces of a specific musical genre.

3.4.1 Individual feature accuracies

The full list of features is the following (numbers correspond to indexes on Figure 3.8):

- 1 to 20: Spectral features (page 105)

- 21 to 33: MFCCs (page 108)

- 34 to 53: First-order differential of spectral features

- 54 to 66: First-order differential of MFCCs

- 67 to 86: Magnitude-normalized first-order differential of spectral features

- 87 to 99: Magnitude-normalized first-order differential of MFCCs

- 100 to 107: Energy in Dixon's subbands (page 111)

- 108 to 115: First-order differential of energy in Dixon's subbands

- 116 to 123: Magnitude-normalized first-order differential of energy in Dixon's subbands

- 124 to 159: Energy in ERB subbands (page 114)

- 160 to 163: Sum of energy in adjacent ERB subbands (page 114)

- 164 to 199: First-order differential of energy in ERB subbands

- 200 to 203: First-order differential of sum of energy in adjacent ERB subbands

- 204 to 239: Magnitude-normalized first-order differential of energy in ERB subbands

- 240 to 243: Magnitude-normalized first-order differential of sum of energy in adjacent ERB subbands

- 244 to 247: Sum in adjacent bands of the magnitude-normalized first-order differential of energy in ERB subbands (i.e. our implementation of (Klapuri et al., 2005) features, see page 114)

- 248 to 256: QMUL features (page 110)

- 257 to 262: Energy in Scheirer's subbands (page 112)

- 263 to 268: First-order differential of the energy in Scheirer's subbands

- 269 to 274: Magnitude-normalized first-order differential of the energy in Scheirer's subbands

Figure 3.8 represents the minimum, maximum and mean accuracy of each feature considered individually with DecisionStump, averaged over all 10 musical genres defined on page 103.

Average relevance On average, the best feature is the first-order differential of the first MFCC (index 54), with an accuracy of 89.1%. The worst feature is the first-order derivative of the spectrum slope (index 51) with an accuracy of 47%; note that the spectrum slope itself (index 18) and its magnitude-normalized first-order differential (index 84) also have low accuracies.

Other individually good features can be found among differentials of spectral features (indexes from 34 to 53 and from 67 to 86), MFCC differentials (though simple first-order differentials seem better than magnitude-normalized ones), many energy features (see dedicated analyses in Section 3.4.3) and high-frequency content differentials (indexes 253 and 255).

Relevance per genre Table 3.5 presents a list of the best and worst features per musical genre. In accordance with the average results above, the first-order derivative of the spectrum slope is the worst feature for each genre. On the other hand, there is

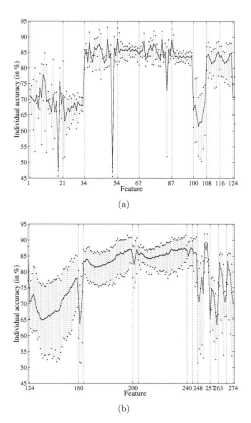

(a)

(b)

Figure 3.8: Individual feature accuracy means and ranges. 3.8(a): Features 1 to 124. 3.8(b): Features 124 to 274.

some variability regarding the best feature. The best feature on average (first-order differential of the first MFCC) is the best on solely two genres out of 10. The first-order derivative of the spectrum mean (index 48) is the best feature on 4 genres. Four other features are best on other genres: the magnitude-normalized first-order derivative of the energy in first ERB subband (index 204), the magnitude-normalized first-order derivative of the spectrum mean (index 81), the magnitude-normalized first-order differential of the first MFCC (index 87) and the first-order differential of the high-frequency content (index 253).

	Best feature		Worst feature	
	Index	Accuracy	Index	Accuracy
Choral	204	85.5%	51	47.7%
Classical	81	87.5%	51	47%
Classical Solo	87	92.5%	51	45.7%
Acoustic	48	87.2%	51	47.2%
Jazz/Blues	48	88.8%	51	47.8%
Balkan/Greek	54	90.2%	51	45.5%
Samba	253	91.7%	51	45.9%
Rock/Pop	54	92.8%	51	47.4%
Afro-American	48	93.1%	51	48.3%
Electronic	48	92.4%	51	47.4%

Table 3.5: Best and worst feature for each genre.

Relevance per musical piece Let us now consider which is the best feature for each musical piece. The analysis shows that a total of 196 features (out of 274), considered individually, are the best feature for (at least) one piece.

3.4.2 Whole feature set accuracy

Using the whole feature set (274 features), NaiveBayes yields an average accuracy of 89.9%, k-NN an average accuracy of 99.5% and C4.5 an average accuracy of 92.5%. Table 3.6 gives accuracies by genre.

These prediction accuracies are very good, especially in the case of k-NN. This method being sensitive to irrelevant features, the very good prediction accuracy achieved with it may indicate that there are not much irrelevant features among those considered. This is an interesting result which confirms what has been found on the preceding page, namely that when considered individually, many features are the best for at least one musical piece.

However, there is certainly a high degree of correlation between some features (consider energy features for instance). These correlations may precisely be the reason for the one NaiveBayes scores worse than the other two methods, as it is indeed sensitive to correlated features.

C4.5 scores in-between the two other methods. Recall that feature selection is embedded in this technique, therefore, for each musical piece, C4.5 prediction accuracy is obtained with less than 274 features. This is an interesting result as it shows that it is possible to score very well without considering the whole feature set.

	NaiveBayes	k-NN	C4.5
Choral	87.7%	**99.8%**	92.9%
Classical	88.1%	99.6%	91%
Classical Solo	87.7%	99.7%	90.7%
Acoustic	85.1%	99%	92.1%
Jazz/Blues	89.7%	99.5%	93.1%
Balkan/Greek	82.9%	99%	88.7%
Samba	86.8%	99.4%	90.9%
Rock/Pop	93.2%	99.7%	94.1%
Afro-American	94%	99.4%	94.6%
Electronic	**94.4%**	99.7%	**94.7%**

Table 3.6: Whole feature set accuracies (274 features) with respect to genres.

Accuracies given in Table 3.6 are averages of musical piece accuracies. Let us now focus on some results for individual pieces: k-NN reaches its minimum accuracy of 86.5% on an *Acoustic* piece, NaiveBayes also on an *Acoustic* piece (54.5%) and C4.5 on a *Jazz* piece (58.5%). On the worst piece for k-NN (86.5%), C4.5 has an accuracy of 89.5%. In fact, there are 12 pieces out of 1360 on which using C4.5 yields a better accuracy than k-NN (2 *Acoustic*, 1 *Electronic*, 5 *Rock/Pop*, 3 *Balkan/Greek* and 1 *Samba*). Therefore, not only is it possible to to score very well with a selection of features instead of the whole feature set but in a few cases, it is also possible to score better.

C4.5 selects a subset of the 274 features for each musical piece, using a particular feature selection technique (see page 96). The selected features may very well change from one piece to the next. Among others, Kohavi and John (1997, Section 2.1) show that decision trees are sensitive to irrelevant features, it would therefore be interesting to discard first those features that are never relevant on any piece and reduce the number of input features to C4.5 by selecting those that have shown relevance on (at least) some pieces. The rest of this chapter precisely addresses this issue in trying to determine if there exists a common subset of features that would perform well for any musical piece. We will come back to this issue in several part of this chapter (3.4.3.3 and 3.4.7).

Prediction accuracies obtained with the whole feature set (best being 99.5%), as well as with individual features (best being 89.1%, see page 118), should be kept in mind when assessing the goodness of any feature set in the remainder of this chapter.

3.4.3 Energy features

Among state-of-the-art beat trackers and beat inducers (Gouyon et al., 2006), many implement a filterbank decomposition of the audio signal and then focus either on frame energy or amplitude in several frequency channels, others focus on differential of energy, in the whole frequency range or in frequency subbands (see Section 2.2.1).

Several frequency decompositions (Sections 3.3.3 and 3.3.4) and two different ways of computing differentials (Section 3.3.5) have been implemented:

1. Energy in the whole frequency range:

   (a) CLAM implementation (page 106): 1 feature

   (b) QMUL implementation (page 110): 1 feature

2. Energy in frequency subbands:

    (a) Energy in Dixon's subbands (page 111): 8 features

    (b) Energy in ERB subbands (page 114): 36 features

    (c) Energy in Scheirer's subbands (page 112): 6 features

3. Sum of energy in adjacent ERB subbands, by groups of 9 (page 114): 4 features

4. Weighted frequency bins:

    (a) Sum of spectral bin magnitude weighted by frequency (page 110): 1 feature

    (b) Sum of spectral bin energy weighted by frequency (page 108): 1 feature

5. Differential of the energy in the whole frequency range:

    (a) First-order differential of CLAM implementation (page 106): 1 feature

    (b) Magnitude-normalized first-order differential[19] of CLAM implementation (page 106): 1 feature

    (c) First-order differential of QMUL implementation (page 110): 1 feature

    (d) Magnitude-normalized first-order differential of QMUL implementation (page 110): 1 feature

6. Differential of the energy in frequency subbands:

    (a) First-order differential of energy in Dixon's subbands (page 111): 8 features (these are our implementation of (Dixon et al., 2003) features)

    (b) Magnitude-normalized first-order differential of energy in Dixon's subbands (page 111): 8 features

    (c) First-order differential of energy in ERB subbands (page 114): 36 features

    (d) Magnitude-normalized first-order differential of energy in ERB subbands (page 114): 36 features

    (e) First-order differential of the energy in Scheirer's subbands (page 112): 6 features (these are our implementation of (Scheirer, 1998) features)

---

[19]The meaning of "magnitude-normalized first-order differential" is explicit on page 116

    (f) Magnitude-normalized first-order differential of the energy in Scheirer's subbands (page 112): 6 features

7. Differential of the sum of energy in adjacent subbands:

    (a) First-order differential of the sum of energy in adjacent ERB subbands (page 114): 4 features

    (b) Magnitude-normalized first-order differential of the sum of energy in adjacent ERB subbands (page 114): 4 features

8. Differential of weighted frequency bins:

    (a) First-order differential of the sum of spectral bin magnitude weighted by frequency (page 110): 1 feature

    (b) Magnitude-normalized first-order differential of the sum of spectral bin magnitude weighted by frequency (page 110): 1 feature

    (c) First-order differential of the sum of spectral bin energy weighted by frequency (page 108): 1 feature

    (d) Magnitude-normalized first-order differential of the sum of spectral bin energy weighted by frequency (page 108): 1 feature

9. Sum in adjacent bands of the magnitude-normalized first-order differential of energy in ERB subbands: 4 features (these are our implementation of (Klapuri et al., 2005) features, as detailed on page 114)

3.4.3.1 Frequency decomposition

Focusing solely on frame magnitudes (leaving differentials aside for the moment), this experiment aims at determining whether it is better to consider the energy on the whole frequency range or to consider a decomposition of the frequency axis in several bands.

Scheirer (1998, p.591), who advocates the latter option, also argues that his algorithm "is not particularly sensitive to particular bands used," that is, that the important point would be to proceed to a frequency decomposition, and not the particular filterbank. However, evaluations performed by Gouyon et al. (2006) suggest that the definition of the frequency filterbank could be a significant issue. Hence, it seems interesting to address the issue of which frequency decomposition is the most appropriate, in case one would be.

This experiment concerns 8 energy feature subsets representing the frequency axis as a whole (1a and 1b), weighted frequency bins (4b and 4a) and decompositions of the frequency axis (2a, 2b, 3 and 2c).[20]

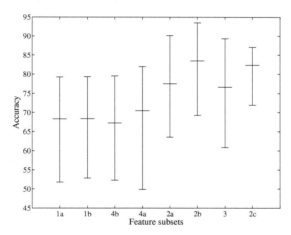

Figure 3.9: Accuracy means and ranges for energy feature subsets.

Figure 3.9 shows the mean accuracy and the range (minimum and maximum accuracy) over the musical genres, according to k-NN (see page 98), for each feature subset. The first conclusion to make from this figure is that, contrarily to Scheirer's argument (1998), there are significant differences between subset accuracies. Subset 2b is the subset with highest mean accuracy and the highest maximum accuracy (93.5% on the *Electronic data set*). Subset 3 is the one with smallest range (i.e. less

---

[20]Note that the number of features differ from one subset to another and ranges from 1 to 36 features.

variance with respect to genres).

All energy subband feature subsets (right of 2a on Figure 3.9) outperform full-band energy feature subsets (left of 2a). This is in accordance with Scheirer's argument (1998), and was already demonstrated (on a much smaller dataset) in (Gouyon and Herrera, 2003a).

The worst accuracy is 49.9%, obtained by subset 4a on the *Choral* data set.

The *Choral* data set is the most difficult for all subsets but subset 2b (for which *Classical* is the most difficult) and subset 2c (for which *ClassicalSolo* is the most difficult). The *Electronic*, *Rock/Pop* and *Afro-American* data sets are the easiest for all subsets.

It should be noted that average accuracies are all worse than those obtained with the whole feature set as well as with the best individual feature, which is not contained in any of these subsets (as a matter of fact, these subset do not account for any of the the individual features that score best for a given genre, as detailed on page 118).

3.4.3.2 Differential vs magnitude

In complement to experiment 3.4.3.1, in this experiment we consider the degree of change of the energy over time, i.e. feature subsets 6a, 6b, 6c, 6d, 7a, 7b, 6e, 6f, 9, 5a, 5b, 5c, 5d, 8c, 8d, 8a and 8b.

We tackle an issue raised in (Gouyon et al., 2006), namely whether energy differential are better features than the mere energy, and incidentally which is the best way to compute differentials.

Figure 3.10 illustrates the mean accuracy and the range (minimum and maximum accuracy, representing the variance with respect to genres), according to k-NN (see page 98), for each feature subset over the musical genres. Subset 6d is the subset with highest mean accuracy (99.4%, similar to that obtained with the whole feature set, obtaining therefore a reduction of feature dimensionality of 7.6) and smallest range (0.8 percent points). Note that this subset is better than any individual feature while it accounts for one of the best individual features, namely that which scores best on the *Choral* data set (see Paragraph 3.4.1).

The worst accuracy is 72.6%, obtained by subset 8c on the *Choral* data set. The

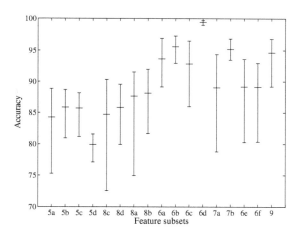

Figure 3.10: Accuracy means and ranges for frequency subband energy differential subsets.

best accuracy is 99.7%, obtained by subset 6d on the *Rock* data set; this is the same accuracy as that obtained when using the whole feature set on this genre (see Table 3.6).

For all subsets, the most difficult data sets are *Choral*, *Classical* and *Acoustic*. Easiest data sets are *Rock*, *Electronic* and *Afro-American*.

Here also, there are significant accuracy differences among feature subsets. Experiment 3.4.3.1 rankings are not exactly preserved but the overall tendency is similar: energy subband feature subsets (right side of Figure 3.10 starting from subset 6a) outperform full-band energy feature subsets (left of subset 6a).

For all energy feature subsets and all data sets, differentials have the clear advantage of increasing the accuracy and reducing the accuracy range around the mean.

For almost all subsets, the magnitude-normalized first-order differential outperforms the mere first-order differential. Exceptions are subset 6e and subset 6f that have similar performances and subset 5c and subset 5d, where normalizing the differential has a negative impact. This is because the energy as computed on page 110 accounts for a logarithm (i.e. a scaling in dB) and the computation of the magnitude-normalized first-order differential also accounts for a logarithm. Computing the logarithm of the logarithm does not make much sense indeed.

Klapuri et al. (2005) argue that "combining [...] adjacent bands [...] is not primarily an issue of computational complexity, but improves the analysis accuracy," i.e. that combining adjacent bands after differentiation would yield better features than individual frequency band differentials. Feature subset 9 implements the former procedure and feature subset 6d implements the latter. As can be seen on Figure 3.10, both perform very well but, contrarily to the argument of Klapuri et al., the latter has a higher average and a much smaller range: feature subset 9 suffers especially on the *Choral* data set, where its accuracy is 89.2% (the second worst accuracy being 93.2%, obtained on the *Classical* data set).

As observed on page 124, there are significant differences in accuracy between different frequency decompositions and it is possible to rank the different frequency decompositions advocated by current state-of-the-art models: the magnitude-normalized first-order differential of the energy in the frequency subbands advocated by Dixon et al. (2003) and Klapuri et al. (2005) show comparable results (with the former having a lower minimum accuracy, on *Choral* data) and are better than subbands proposed by Scheirer (1998).

3.4.3.3 Frequency subband rankings

Given a particular decomposition in frequency subbands, one may wonder whether energies in some particular subbands are, on average, more representative of beats than some other subbands and may therefore be given relatively more importance in a computational system. For instance, in the case of algorithms that compute periodicity functions over several subband energy values, or amplitude envelopes, or their sum thereof, one may wonder whether some bands should be given a particular weight *a priori*. some frequency regions may even be discarded and an algorithm may focus solely on a restricted region of the spectrum. Indeed, some authors advocate that a particular emphasis should be put on low frequencies (Blum et al., 1999; Alghoniemy and Tewfik, 1999) or equivalently low-pitched MIDI events (Dixon and Cambouropoul

2000), while others argue that energy content at high frequencies should be emphasized (Laroche, 2003) or that both low and high frequencies should be given more
importance than mid-range frequencies (Heittola and Klapuri, 2002).

For this experiment, we focus on energy feature subsets 2b, 6c and 6d as they
correspond to the frequency decomposition with the greater number of bands (and
are therefore more indicated for illustration purposes). For each of these 3 sets, we
consider for each musical piece the worth of different subsets, automatically generated
and evaluated with the method described on page 99. For each feature of the particular set considered, we then make a histogram of the number of pieces over which it is
part of the winning subset. Histograms for two feature sets are shown in Figure 3.11.
Feature subset 6c is not represented, but shows similar trends than subset 2b.

Figure 3.11(a) shows that energy in both low and high frequencies (approximately
from band 1 to 5, i.e. between 50 and 300 Hz, and from band 27 to 36, i.e. between 5.5
and 20 kHz) are more relevant than energy in mid-range frequencies when considering
energy magnitude (or first-order differential). On the other hand, Figure 3.11(b)
suggests that solely low frequencies (roughly between band 1 and 7, i.e. between 50
and 500 Hz) are more relevant than others when considering magnitude-normalized
first-order differentials of energy.

Detailed analyses for each musical genre shows the same "U-shape" as that observed on Figure 3.11(a) for the 3 feature subsets on many genres. A different trend
is however typical of some genres: Subset 6d shows linear trend, where relevance is
inversely proportional to frequency (low frequencies being therefore relatively more
relevant), for the following genres: *Balkan/Greek, Samba, Choral, Classical, Classical
Solo* and *Jazz/Blues*. Subsets 2b and 6c show the same linear trend solely on the
Classical and *Classical Solo* data sets.[21]

It is informative at this point to compare these results with individual feature
relevance results, as shown on Figure 3.8, on page 119. On this figure, one can
clearly see similar U-shapes in most of the energy feature subsets (between indexes
100 and 107, 108 and 115, 116 and 123, 124 and 159, 160 and 163, 164 and 199,

---

[21]Note however that for some genres, it is difficult to make solid conclusions as they are not
represented by many musical pieces (e.g. *Choral*).

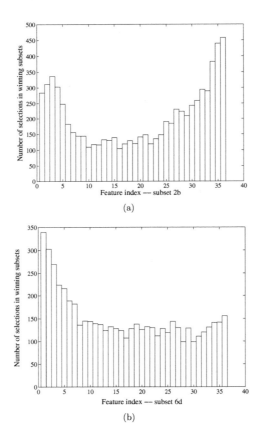

Figure 3.11: Frequency subbands ranking for subset 2b (Figure 3.11(a)) and subset 6d (Figure 3.11(b))

200 and 203, 204 and 239 and finally 240 and 243), indicating that, for whatever frequency decomposition, mid-range frequency bands are less relevant than low and high frequency bands.

From all these informations, it seems reasonable to conclude that frequency bands roughly between 500 to 5000 Hz are relatively less representative of beats than low frequencies (below 500 Hz) and high frequencies (above 5000 Hz).

One may therefore wonder whether this means that we should discard mid-range frequency bands. Considering solely bands 1 to 7 and 27 to 36, the average accuracy of energy differentials (i.e. subset 6c) is 93.1%, the minimum accuracy is 87.1% and the maximum accuracy is 95.9% (this should be compared to accuracies reached when considering all features of subset 6c: 92.8%, 85.9% and 96.4% respectively). Considering now magnitude-normalized differentials (subset 6d), the average accuracy obtained when using solely these low and high frequency regions is 98.6%, the minimum accuracy is 97.6% and the maximum accuracy is 99.2% while these accuracies were respectively 99.4%, 98.9% and 99.7% when using all subbands of subset 6d. We can see that focusing solely on subbands below 500 Hz and above 5 kHz (and obtaining a dimensionality reduction of a factor of 2.1) the results are comparable to those obtained with all the 36 bands. Further, we obtain better results with this manual selection of features than with the automatic feature selection technique embedded in C4.5 (using C4.5 and subset 6d yields average, minimum and maximum accuracies of 92.4%, 88.5% and 94.9%, respectively). Accounting for mid-range frequencies does not damage results, a conservative implementation may therefore also consider them. But in an implementation which could not afford to consider many subbands, these should be the first candidates for being discarded.

3.4.4 Spectral features

Let us now focus on the spectral features whose implementations are detailed in Section 3.3.1. (Note that we do not include energy features in this experiment as they are the object of previous experiments.) They are divided in 3 subsets:

1. Spectral feature magnitudes:

(a) Spectral peak first tristimulus

(b) Spectral peak harmonic centroid

(c) Spectral peak harmonic deviation

(d) Spectral peak mean

(e) Spectral peak second tristimulus

(f) Spectral peak third tristimulus

(g) Centroid

(h) Flatness

(i) Geometric mean

(j) Magnitude kurtosis

(k) Low-frequency energy relation

(l) Maximum magnitude frequency

(m) Mean

(n) Roloff

(o) Magnitude Skewness

(p) Slope

(q) Spread

(r) ZCR

2. Spectral feature first-order differentials (a—r)[22]

3. Spectral feature magnitude-normalized first-order differentials (a—r)

Figure 3.12 provides a comparison of the performances of the 3 feature subsets. According to k-NN, average accuracies over genres are 93.1% for subset 1, 98.4% for subset 2 and 98.9% for subset 3. Accuracy ranges are respectively 5.1, 1.6 and 0.8

---

[22]For the sake of simplicity, we do not provide the whole list here; differentials have the same ordering as features 1a to 1r. For instance, feature 2e refers to the first-order differential of the spectral peak second tristimulus.

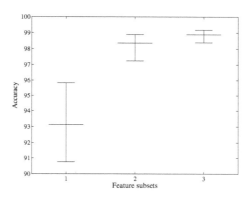

Figure 3.12: Accuracy means and ranges for spectral feature subsets.

percent points. For all subsets, the worst performance is obtained on the *Acoustic* data set.

We saw in Section 3.4.2 that the accuracy of the whole feature set is 99.5% with k-NN. Using solely the magnitude-normalised differentials of spectral features (i.e. lowering the number of features by a factor of 15.2) yield a comparable result. This feature set, who accounts for one of the best individual features but does *not* account for energy features, are therefore very relevant.

One can see that, as for the energy feature subsets (experiment 3.4.3.2), differentials have the clear advantage over frame magnitudes of raising the average accuracy and reducing the accuracy range between genres. Further, here also, the magnitude-normalized first-order differential performs slightly better than the mere first-order differential.

The worth of different subsets of feature subset 3 has been considered for each musical piece (subsets being automatically generated and evaluated with the method described on page 99), Figure 3.13 illustrates a histogram of the number of pieces over which each feature of subset 3 is a member of the winning subset. This ranking of features can be compared to that shown in Figure 3.8 (between index 67 and index 86), where features were considered individually, it has approximately the

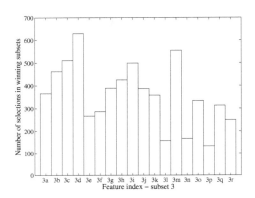

Figure 3.13: Spectral features rankings, feature subset 3

same shape (recall that index 74 and 77 in Figure 3.8 represent energy features that are not considered in Figure 3.13). Accordingly to previous results, feature 3p (the magnitude-normalized differential of the spectrum slope) is the worst feature in this subset. Depending on the dimensionality reduction desired, it is possible to define a threshold at some height on Figure 3.13 in order to select the most relevant features. For instance, if one wants to keep 6 spectral features, the best are 3b, 3c, 3d, 3h, 3i and 3m, that is, the magnitude-normalized differential of the following features: the spectral peak harmonic centroid, the spectral peak harmonic deviation, the spectral peak mean, the spectrum flatness, the spectrum geometric mean and the spectrum mean.

An analysis of feature relevance per genre shows that in addition to the 6 previous features, features 3j and 3o (magnitude-normalized differentials of the spectrum magnitude kurtosis and skewness) are relevant for *Choral* data and feature 3k (the magnitude-normalized differential of the low-frequency energy relation) is relevant for *Jazz/Blues* and *Balkan/Greek* data.

It is interesting to remark that rhythm description literature accounts for very few references to spectral features, with the exception of (Goto and Muraoka, 1999), (Seppänen, 2001), (Gouyon and Herrera, 2003b) and (Sethares et al., 2005), and that these references do not consider the features selected here.

3.4.5 Cepstral features

Let us consider now MFCCs (page 108) and define the following feature subsets:

1. MFCCs, except the first coefficient (12 features)

2. First-order differential of MFCCs, except the first coefficient (12 features)

3. Magnitude-normalized first-order differential of MFCCs, except the first coefficient (12 features)

We have seen in the part related to individual feature accuracies (Section 3.4.1) that the first-order derivative of the first MFCC is a very relevant feature. However, this MFCC is a measure of the signal energy, therefore we do not consider it in this experiment.

Figure 3.14 illustrates the minimum, maximum and mean accuracies of the three feature subsets on the whole data set.

According to k-NN, average accuracies over genres are 87.1% for subset 1, 98.7% for subset 2 and 98.4% for subset 3. According to genres, accuracy ranges are respectively 10.4, 5.2 and 1.6 percent points. Accuracies of subsets 2 and 3 are comparable with that obtained with the whole feature set (99.4%). These features are therefore very relevant. Figure 3.8(a) (indexes 54 to 66 and 87 to 99) also illustrates that they are all relevant (with approximately the same relevance) when considered individually.

Subset 1 obtains its worst performance on the *Choral* and *Classical* data sets and its best performances on the *Samba*, *Rock* and *Electronic* data sets. In between, the distribution of accuracies is approximately uniform. Subsets 2 and 3 behave differently, they both show similar accuracies for all genres (at the exception of an outlier for subset 2, the *Choral* data set —91.6%—; considering the second minimum instead of it, the range raises from 5.2 to 1.8 percent points).

As for the energy feature subsets (experiment 3.4.3.2), differentials have the clear advantage over frame magnitudes of raising the average accuracy and reducing the

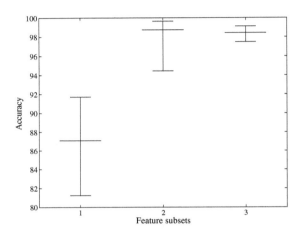

Figure 3.14: Accuracy means and ranges for MFCC feature subsets.

accuracy range between genres. However, unlike for energy features, the first-order differential seems better than the magnitude-normalized first-order differential.

3.4.6 Onset detection functions

Here, the features of interest are the following:[23] (Note that, as in Sections 3.4.4 and 3.4.5, we do not include energy features in this experiment as they are the object of previous experiments.)

1. Complex spectral difference

2. Phase deviation

3. Spectral difference

When using k-NN, the average accuracy over musical genres is 82.8%, and the range is 30.7 percent points. The worst performance is obtained on *Choral* data (60.5%) and the best performance is obtained on *Electronic* data (91.2%).

---

[23]see page 110 for details

Despite the fact that this performance is worse than those already obtained with the whole feature set, energy features, spectral features or MFCCs, this performance is interesting as it is better than any of the individual performance of these 3 features (see Figure 3.8(a), indexes 250 to 252, the best individual feature being the phase deviation).

3.4.7 Combining features

The experiments described above focused on "families" of features (as energy features for instance). Rankings were obtained between families and features of a given family. Let us now consider combining different families of features. Selecting the features that performed best yields the following subset of 59 features:

- magnitude-normalized first-order differentials of the energy in ERB bands (36 bands),

- magnitude-normalized first-order differentials of the spectral peak harmonic centroid, spectral peak harmonic deviation, spectral peak mean, flatness, geometric mean, mean, kurtosis, skewness and low-frequency energy relation (9 features)

- the first-order differentials of the MFCCs (13 coefficients, including MFCC1)

- phase deviation

Using this subset, k-NN yields an average accuracy of 99.6% (and minimum and maximum accuracies of 99.4% (obtained on *Acoustic* data) and 99.9% (obtained on *Choral* data) respectively).

3.4.8 "All instances in a bag" approach

As mentioned in this chapter introduction, other methodologies than ours could be used to seek the best discriminating features between instances of the two classes "beat" and "non-beat." For instance, instead of seeking feature relevance on individual musical pieces and averaging the results, Seppänen (2001) considers all instances at once and seeks the features that yield the best beat and non-beat *models*.

We preferred a "musical piece based" approach over an "all instances in a bag" approach for several reasons. First, we foresaw that different sets of features would probably be relevant depending on the musical style. And it is easier to keep track of this information when considering pieces individually. Second, and more importantly, our intuition led us to doubt of the existence of such a concept as a beat model. Intuitively, it is hard to believe that specifications of a set of feature *values* would be representative of beats on *any* piece of music. Indeed, a specific feature may have recurrent values on beats of several musical pieces but the actual (recurrent) value may very well differ from one piece to the other, consider for instance the differences between a piece whose volume has been enhanced and its original version. Our approach is rather qualitative than quantitative as we intend to determine *which* are the relevant features to focus on and which are the features to discard, rather than values for such features.

However, we also conducted a few experiments on all instances considered at once. Five beat and five non-beat instances have been randomly selected from each musical piece, yielding a total of 13598 instances.[24] Half of these instances have been considered as training instances (6800 instances) and the other half as testing instances (6798 instances); special care has been taken so that both sets cannot contain beats and non-beats of the same musical piece.

Beat and non-beat models have been designed with the training dataset and evaluated with the testing dataset. Accuracies in the following paragraphs are accuracies obtained on the testing data and do not entail cross-validations as in previous experiments.

All features   Considering the whole feature set and k-NN as induction algorithm, additional resampling of the training and testing data to 25% of the original number of instances is necessary for computational reasons (i.e. 1700 beat instances and 1699

---

[24]On 2 specific pieces, the number of non-beats was restricted to 4.

non-beat instances remain).

The accuracy reached is 99.7%. This should be compared to the average accuracy over all musical pieces considered individually (see on page 120): 99.5%.

Best features Considering now solely the 59 features that were scored best on previous experiments (listed in Section 3.4.7), the accuracy is comparable: 99.3%. (Additional resampling was not necessary here.) Average accuracy over all musical pieces considered individually was comparable (see on page 137): 99.6%.

Individual feature accuracies Let us now consider whether some single feature(s) among those 59 features produce comparable results when considered individually, or whether we really need a combination of them.

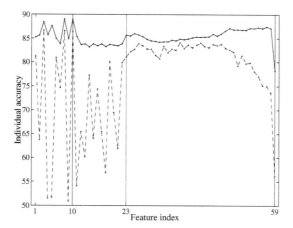

Figure 3.15: Individual accuracies of the best features. The plain line represents average accuracies over all musical pieces (see Figure 3.8), the dashed line represents the accuracies obtained when grouping the instances of all pieces.

Figure 3.15 illustrates individual accuracies of the 59 best features. (Indexes 1 to 9 represent the selected spectral features, indexes 10 to 22 the selected cepstral features, indexes 23 to 58 the selected ERB-based features and index 59 the phase deviation.) The dashed line represents accuracies obtained when considering all instances at once while the plain line represents average accuracies obtained on individual musical pieces.

We can observe that no single feature reaches the 99.3% accuracy obtained when considering the 59 features jointly. All features are relevant and combination of them is necessary to obtain the best results.

We can also observe that for any feature among these 59 features, the average accuracy over musical pieces is always better than the accuracy obtained over all instances considered jointly. In our opinion, this is an argument in favor of the "musical piece based" approach over the "all instances in a bag" approach. One may object that the former method probably induces an overfitting to the data (beat models are very specialized on particular pieces) that the latter would rule out. However, as we already mentionned, our purpose is not to seek good generalization of beat models, but rather good description of individual pieces, then the process of averaging over many pieces guarantees that features that are good only on very particular cases are ruled out.

3.5   Conclusions

In this chapter, we addressed the determination of the low-level features of musical audio signal that convey best the rhythmic aspects of musical signals.

The best individual feature is the first-order differential of the first MFCC, which amounts to the signal's variation of energy. This is not surprising as the energy differential is correlated with note onsets, and note onsets are long thought to be of prime importance in rhythm description. Many existing rhythm description systems are based on a front-end that computes the energy or its differential. When considered individually, the following features are more accurate than any other on a specific musical genre (among the 10 genres considered here): first-order derivative and magnitude-normalized first-order derivative of the spectrum mean, magnitude-normalized first-order derivative of the energy in first ERB subband, magnitude-normalized first-order differential of the first MFCC, and the first-order differential of the high-frequency content.

However, an increase of the accuracy is achieved when combining features. Results using the whole feature set show that many features are relevant but also that many are correlated.

In accordance with a point raised by Scheirer (1998), it is shown that a decomposition of the frequency axis in several bands provides better features than the energy computed on the whole frequency range. However, unlike Scheirer proposes, the definition of the frequency decomposition does have a significant impact, an ERB filterbank being the best choice among those tested here. Then, as Gouyon et al. (2006) propose, for all energy feature subsets, the degree of change of feature magnitudes are better features than mere magnitudes. Normalizing a first-order magnitude differential by the magnitude seems to be the most accurate implementation for that (it performs better than the mere differential). Another conclusion is that individual frequency band differentials seem to be better than combination of adjacent bands into fewer features (contrarily to a point raised in (Klapuri et al., 2005)). Another conclusion regarding energy values in frequency subbands is that some frequency regions are more relevant than others, namely low and high frequencies (approximatively below 500 Hz and above 5 kHz).

Spectral features are also very relevant, but they score differently, the worst is the spectrum slope (which should be discarded) and the best are the spectral peak harmonic centroid, the spectral peak harmonic deviation, the spectral peak mean, the spectrum flatness, the spectrum geometric mean, the spectrum mean, the spectrum magnitude kurtosis and skewness and the low-frequency energy relation. Here also, computing the magnitude-normalized first-order differential results in an increase of the accuracy.

Very good accuracy figures can also be achieved with MFCC differentials. This is interesting as, to our knowledge, there exist no references to the use of such features in rhythm description tasks. Unlike for other features, the mere first-order differential outperforms the magnitude-normalized first-order differential. This can be explained by the fact that an MFCC is computed as $DCT(log(x_t))$, where DCT is the Discrete Cosine Transform and x_t represents the power values of the signal in a specific Mel band (see on page 108). Hence, the differential of an MFCC can be written as $\frac{d}{dt}(DCT(log(x_t)))$. As DCT and $\frac{d}{dt}$ are linear operations, this can be rewritten as $DCT\left(\frac{d}{dt}(log(x_t))\right)$, which is equivalent to $DCT\left(\frac{\frac{d}{dt}x_t}{x_t}\right)$. Hence we see that, as the MFCC computation entails a logarithm, the computation of its differential is actually already equivalent to a *magnitude-normalized* first-order differential.

We propose the following set of 59 features as the best feature set: the magnitude-normalized first-order differentials of the energy in ERB bands; the magnitude-normalized first-order differentials of the spectral peak harmonic centroid, spectral peak harmonic deviation, spectral peak mean, flatness, geometric mean, mean, kurtosis, skewness and low-frequency energy relation; the first-order differentials of the MFCCs and the phase deviation.

Chapter 4

Selected features in context — Tempo induction

In Chapter 3, we made the assumption that low-level audio features whose temporal behaviors reflect beat and non-beat positions (as determined by classification experiments) would also be promising features for the computation of periodicity functions. Consequently, they should also be useful features for inducing tempo. One objective of this chapter is to illustrate the relevance of some feature sets selected in Chapter 3 in an actual tempo induction task. A number of open issues in tempo induction research were raised on page 82 in addition to the issue of low-level feature selection. Some are addressed in this chapter, namely, whether a particular periodicity function is better than other ones, what is the best way to combine and to parse multiple information sources (e.g. integration before or after periodicity detection) and whether the joint estimation of several metrical levels helps the determination of tempo.

In this chapter, different combinations of feature sets and tempo induction algorithms are presented and discussed with the following organization: Section 4.1 details the implementation of the tempo induction algorithms. Section 4.2 covers experiments conducted with these diverse algorithms and different feature sets. Section 4.3 provides a comparison with some state-of-the-art algorithms. Finally, Section 4.4 provides conclusions regarding the best algorithm/features combination.

4.1 Tempo induction algorithms

4.1.1 Features

In the rest of this chapter, we consider the following feature sets:

1. the first-order differential of MFCC1: 1 feature

2. the magnitude-normalized first-order differential of MFCC1: 1 feature

3. the magnitude-normalized first-order differential of energy in Dixon's subbands: 8 features

4. the energy in ERB subbands: 36 features

5. the first-order differential of energy in ERB subbands: 36 features

6. the magnitude-normalized first-order differential of energy in ERB subbands: 36 features

7. the sum in adjacent bands of the magnitude-normalized first-order differential of energy in ERB subbands (i.e. our implementation of the features proposed by Klapuri et al. (2005)): 4 features

8. the first-order differential of MFCCs: 13 features

9. the magnitude-normalized first-order differential of MFCCs: 13 features

10. the magnitude-normalized first-order differential of the 9 best spectral features (see page 131): 9 features

11. the magnitude-normalized first-order differential of energy in Dixon's subbands 1, 2, 7 and 8: 4 features

12. the magnitude-normalized first-order differential of energy in Dixon's subbands 3, 4, 5 and 6: 4 features

13. the feature set selected on page 137 in Chapter 3: 59 features

14. feature set 13, with the exception of the magnitude-normalized first-order differential of energy in ERB subbands (only one energy feature is left, the first-order differential of MFCC1): 23 features

4.1.2 Periodicity function computation

We present here the implementation of three periodicity functions among the many possible ones listed in Paragraph 2.2.2.2.

4.1.2.1 Autocorrelation

The autocorrelation function (ACF) $r(\tau)$ of a discrete signal $x(n)$ is computed as follows:

$$r(\tau) = \sum_{n=0}^{N-\tau-1} x(n)x(n+m) \qquad \forall \tau \in \{0 \cdots U\}$$

where N is the number of samples of the signal and U is the upper limit for the lag τ.[1]

We used the Matlab implementation of the ACF that normalizes the function so that $r(0) = 1$ and used a maximum lag (upper limit U) of 5 s.

4.1.2.2 Fourier transform

The Fourier transform converts a time domain signal $x(t)$ into its frequency spectrum $X(f)$. The definition of the Fourier transform is the following:

$$X(f) = \int_{-\infty}^{\infty} x(t)e^{-i2\pi ft}dt$$

where t is the continuous time index (in seconds) and f the continuous frequency index (in Hertz).

As we are dealing with sampled and finite signals, we make use of the Discrete

---

[1] In this definition of the ACF, the integration time is set to the maximum $(N - \tau)$ given the length of the signal. Note that it could be set to smaller values (Brown, 1993).

Fourier Transform (DFT), whose definition is the following:

$$X(k) = \sum_{n=0}^{N-1} x(nT_S)e^{-i2\pi knT_S}$$

where T_S is the sampling period, N is the number of samples of the signal, n is the discrete time index and k is the discrete frequency index.

We used the Matlab Fast Fourier Transform algorithm for computing the DFT of feature lists.

4.1.2.3   Comb filterbank

We implemented a filterbank of constant half-time comb filter resonators as proposed by Scheirer (1998). Details of the implementation are based on (Klapuri et al., 2005). Given an input signal $x(t)$, the output of a comb filter with delay τ and gain α_τ is

$$y_\tau(t) = \alpha_\tau y_\tau(t - \tau) + (1 - \alpha_\tau)x(t)$$

Comb filters have an exponentially-decaying impulse response where the half-time refers to the delay required for the response to decay to half its initial value. The gain α_τ of a filter depends on the filter's delay τ. As we want the filters to have equivalent half-time, α_τ is set differently for each filter as $\alpha_\tau = 0.5^{\tau/t_0}$, where t_0 is the half-time. We chose a half-time of 2 s (instead of 3 s proposed in (Klapuri et al., 2005)). As Klapuri et al. (2005, equation 7) propose, the instantaneous energies of each comb filter are computed as

$$z_\tau(t) = \frac{1}{\tau} \sum_{i=t-\tau+1}^{t} (y_\tau(i))^2$$

They are then normalized to obtain

$$s_\tau(t) = \frac{1}{1 - \gamma(\alpha_\tau)}(\frac{z_\tau(t)}{w(t)} - \gamma(\alpha_\tau))$$

where $w(t)$ is the energy of the signal $x(t)$, calculated by squaring $x(t)$ and applying a resonator with $\tau = 1$ (a leaky integrator) and $\gamma(\alpha_\tau)$ is the overall power of a comb filter with gain α_τ. See (Klapuri et al., 2005) for more details. Given the feature sampling rate used here (86.1 Hz), a total number of 85 resonators is needed to cover the tempo region between 50 and 250 BPM. The 85 temporal functions $s_\tau(t)$ are integrated over time to yield the final periodicity function.

4.1.3 Combining and parsing multiple information sources

We saw on page 46 that the rationale for the integration of diverse information sources is an important design choice in a tempo induction algorithm: either *before* or *after* to the computation of periodicity functions from low-level features. Another design choice stands in the choice of the integration operation (e.g. sum, product), and other ones stand in the consideration of a feature evaluation criterion or a normalization of the features.

In the experiments detailed in Section 4.2, we consider the following different combinations:

- Normalizing features, summing or multiplying them, computing a periodicity function on the resulting function and picking the maximum peak in a specific tempo region (in the following experiments, we focus on the region between 50 and 250 BPM).

- Summing, or multiplying, the periodicity functions of several features and picking the maximum peak (e.g. between 50 and 250 BPM).

- Computing a periodicity function for each feature, selecting peaks in each periodicity function and parsing the peak list accounting explicitly for influential schemes between metrical levels, as proposed by Dixon et al. (2003) and detailed on page 62 (this method is later referred to as "musical parsing").

Different periodicity functions are tested for each combination, namely the ACF, comb filterbank and Fourier transform.

Note that other parsing methods could also have been considered, as for instance seeking periodicities in periodicity functions themselves, as proposed by Gouyon et al. (2002), or keep several candidates (prominent peaks in the periodicity function) and refine them through beat tracking, as proposed e.g. by Dixon (2001a). Exhaustive evaluations of these methods are left for future work.

In summary, in the rest of this chapter, we will refer to the following algorithms in association with the diverse feature sets on page 144:

Algorithm-1 Normalizing and summing features, computing an ACF of the resulting function and picking the maximum peak (this corresponds to the second column in Table 4.2).

Algorithm-2 Normalizing and summing features, using a comb filterbank on the resulting function and picking the maximum peak (third column in Table 4.2).

Algorithm-3 Normalizing and summing features, computing a Fourier transform of the resulting function and picking the maximum peak (fourth column in Table 4.2).

Algorithm-4 Computing an ACF for each feature, summing these functions and picking the maximum peak of the resulting function (fifth column in Table 4.2).

Algorithm-5 Using a comb filterbank for each feature, summing the functions and picking the maximum peak of the resulting function (sixth column in Table 4.2).

Algorithm-6 Computing a Fourier transform for each feature, summing these functions and picking the maximum peak of the resulting function (seventh column in Table 4.2).

Algorithm-7 Computing an ACF for each feature, multiplying these functions and picking the maximum peak of the resulting function (eighth column in Table 4.2).

Algorithm-8 Using a comb filterbank for each feature, multiplying the functions and picking the maximum peak of the resulting function (ninth column in Table 4.2).

Algorithm-9 Computing a Fourier transform for each feature, multiplying these functions and picking the maximum peak of the resulting function (tenth column in Table 4.2).

Algorithm-10 Normalizing and multiplying features, computing an ACF of the resulting function and picking the maximum peak.

Algorithm-11 Normalizing and multiplying features, computing a Fourier transform of the resulting function and picking the maximum peak.

Algorithm-12 Computing an ACF for each feature, applying the parsing and integration method proposed by Dixon et al. (2003) (see description on page 62, it corresponds to the eleventh column in Table 4.2)

Algorithm-13 Same as Algorithm 12, but using a comb filterbank instead of the ACF (twelfth column in Table 4.2).

Algorithm-14 Same as Algorithm 12, but using the Fourier transform instead of the ACF (thirteenth column in Table 4.2).

4.2 Experiments

4.2.1 Data

As in Chapter 3, the data used in this chapter comes from different sources (personal collections, publicly-available data and commercial sound libraries) and comes with different types of legacy metadata.[2] There is a total of 3223 audio files.

---

[2]Here also, we wish to thank those who helped in the recollection of this data and metadata: Miguel Alonso, The Tape Gallery, Pedro Cano and Simon Dixon.

Loops   This data set is made up of 2036 files, it is described on page 66. It has been used in (Gouyon et al., 2006) in the ISMIR 2004 tempo induction contest See page 66 for details on annotations and availability of this data. Figure 2.6(c) illustrates the distribution of the excerpts along the tempo axis.

Ballroom   This data set is made up of 698 files, it is described on page 66. It has been used in (Gouyon et al., 2006) in the ISMIR 2004 tempo induction contest and in (Gouyon et al., 2004a; Gouyon and Dixon, 2004; Dixon et al., 2004) in rhythm classification experiments. See page 66 for details on annotations and availability of this data. Figure 2.6(b) illustrates the distribution of the excerpts along the tempo axis.

Alonso   This data set is made up of 489 files in .wav format, ripped from commercial CDs. These files were given to us by Miguel Alonso[3] with a sampling rate of 16 kHz, and were resampled to 44100 Hz with Sox. File length are around 20 s each and range from 50 BPM to 200 BPM. Figure 4.1 illustrates the distribution of the excerpts along the tempo axis. Some have varying tempo. This data set has been used by Alonso et al. (2004) in tempo induction experiments. It comes with beat annotations at one metrical level and genre annotations in 10 classes. We did not cross-checked the tempo annotations. The genre distribution is detailed in Table 4.1. Audio data is not publicly available for copyright reasons.

4.2.2   Evaluation metrics

Two evaluation metrics are used in this chapter:

- *Accuracy 1* (acc1): The percentage of tempo estimates within 5% of the ground-truth tempo.

- *Accuracy 2* (acc2): The percentage of tempo estimates within 5% of either the ground-truth tempo, or half, twice, three times or one-third of the ground-truth tempo.

---

[3]from the École Nationale Supérieure des Télécommunications (ENST) in Paris

Genre	# pieces
Classical	137
Electronic/Dance	23
Jazz	79
Latin	37
Miscellaneous	55
Pop	40
Rap/Hip-Hop/Trip-Hop	20
Reggae	30
Rock	44
Soul	24

Table 4.1: Genre distribution of Alonso data set

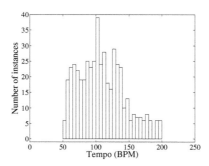

Figure 4.1: Histograms of Alonso's data ground-truth tempo values in 5 BPM steps.

They are basically the same accuracy measures as those used in Section 2.3, page 70, except the precision window is slightly wider (5% instead of 4%).

4.2.3 Results

Results (with respect to accuracy 2) of different algorithms and feature sets associations on Alonso data are reported in Table 4.2. The performances of some algorithms are given in more details (i.e. both accuracy measures and the rest of the test data set) in Appendix A on page 237.

We do not report on results obtained by algorithms 10 and 11 (normalizing features, *multiplying* them (instead of summing), computing a periodicity function on the resulting function and picking the maximum peak). Indeed, the performances obtained with these were extremely low for both periodicity functions.

In this section we focus on performances on Alonso data only (apart when explicitly stated) in order to keep a relatively balanced representation of timbral characteristics and musical genres in the test data. The Loops and Ballroom data sets may induce a bias towards specific timbres or genres. They will nevertheless be useful for some experiments such as the comparison with state-of-the-art algorithms.

4.2.3.1 Feature sets

A first observation is that feature set rankings differ with respect to the tempo induction algorithms they are combined with. As one can see highlighted in bold fonts in Table 4.2, deciding which is the best feature set depends on the algorithm. Only 6 feature sets (out of 14) are never the best for any algorithm. On the other hand, the worst feature set is the same for all algorithms: set 9, the magnitude-normalized first-order differential of MFCCs.

Energy in frequency subbands   According to Chapter 3, on page 122, energy computed in several frequency bands should provide better features than the energy computed on the whole frequency range. This seems to be verified here, indeed, when used as input of most algorithms, feature sets 1 and 2 score worse than feature sets 3, 6 or 7.

Also according to Chapter 3, on page 124, using energy values (or differential thereof) in a particular subband decompositions or another should have a significant repercussion on the performances. Different subband decompositions should score differently. We can verify that, when combined with Algorithm 12, 5 and 14, the performances of sets 3, 6 and 7 differ significantly. This is however not true for all algorithms. For specific periodicity functions and combining and parsing strategies (summing the Fourier transforms of the different features and selecting the maximum peak for instance, i.e. Algorithm 6), the performances of these feature sets are similar.

According to findings of Chapter 3, the 36 individual magnitude-normalized first-order differential of energy in ERB subbands (feature set 6) should be better than their sum in 4 adjacent bands (feature set 7). (For illustration, see the performances of feature sets 6d and 9, respectively, on Figure 3.10 on page 127.) Here, feature set 6 clearly outperforms feature set 7 only when combined with Algorithm 12. For other algorithms, either these two sets show similar accuracies or the opposite is true (e.g. with Algorithm 14).

Further, the ranking of subband decomposition motivated in Chapter 3 on page 126 (that is, that set 6 should be better than set 3, which in turn should be better than set 7) is not always respected here. For example, set 3 outperforms the two other sets when combined with Algorithm 12 and Algorithm 13.

Another finding of Chapter 3 is that magnitude-normalized first-order differential of feature values should be better than first-order differential, which in turn should be better than mere feature values (see on page 126). When considering the energy in ERB bands (feature sets 6, 5 and 4 respectively), this is confirmed for some algorithms (e.g. Algorithm 3), but not for some others (e.g. Algorithm 12).

One last conclusion of Chapter 3 with respect to energy in frequency subbands concerns their relative relevance: energy in low and high frequency bands should be better than in mid-frequency bands (see on page 128). This is verified here, indeed, when used as input to any algorithm, feature set 11 scores better than feature set 12.

Spectral features and MFCCs According to Chapter 3, relatively good accuracy figures should be obtained with the set of 9 spectral features promoted on page 131, as well as with the first-order differential of MFCCs. The results of feature set 10 and 8 confirm this here. Feature set 8 is even the best set for Algorithm 6.

These experiments also confirm a point raised on page 135: unlike for other features, the first-order differential of MFCCs (feature set 8) are always better than the magnitude-normalized first-order differential (feature set 9). This is also true when considering MFCC1 alone: set 1 is always better than set 2.

"Best" feature set   Chapter 3 (page 137) concluded on the special relevance of a set of 59 features, combining energy features, spectral features and MFCCs (set 13). We can see in Table 4.2 that for many algorithms, there exists feature sets that perform better than this set. However, this feature set shows accuracies close to the best feature set when combined with all algorithms and it is the best set when combined with Algorithm 9. Looking more into details (see Appendix A), we can also see that it is the best set for the Ballroom data set when associated with Algorithm 12, for the Loops data set and the overall set when associated with Algorithm 1 or Algorithm 4 and on the overall data set when associated with Algorithm 9.

4.2.3.2   Periodicity functions

The results of Table 4.2 show that the use of a comb filterbank as periodicity function provides in most cases better results than that of the autocorrelation, which in turn is better than the Fourier transform.

Results on the whole data set (and not solely Alonso data) presented in Appendix A also show that the use of the autocorrelation generally provides better results than the Fourier transform.

4.2.3.3   Combining and parsing multiple information sources

Sum vs product   The product of the feature lists before periodicity detection is harmful in all the cases (Algorithm 10 and Algorithm 11), and leads to performances always below 5%, therefore we will not discuss it further.

A comparison of the performances of algorithms implementing sums or products of periodicity functions (Algorithms 4 and 7 for the ACF, Algorithms 6 and 9 for the Fourier transform and Algorithms 5 and 8 for the comb filterbank) does not reveal that a specific method would be better than the other one, this seems to depend on the input feature set.

Integration before or after periodicity detection   The performances of Algorithm 4 (implementing the sum of ACFs) are slightly better than those of Algorithm 1 (implementing the ACF of the sum of feature lists) for the majority of feature sets.

On the other hand, no clear trend can be seen when using the Fourier transform nor a comb filterbank instead of the ACF.

Additivity   In this paragraph, we address the question of the performance of combined feature sets and focus on feature set 13 (selected in Chapter 3 as the most representative of beats), which is the combination of feature sets 6 and 14. We focus on experiments conducted on the whole data set, whose full results are given in Appendix A.

An analysis of errors made by feature sets 6 and 14 when combined with Algorithm 12 shows that 30.7% of the musical pieces (i.e. 161 pieces) whose tempo has been wrongly computed when using feature set 6 have been correctly processed when using feature set 14; on the other hand, 26.9% of the musical pieces (i.e. 134 pieces) whose tempo has been badly computed when using feature set 14 have been correctly processed when using feature set 6 and 364 pieces are badly computed by both. When combined with Algorithm 1, these figures become respectively 28.2% and 35.4%, 851 pieces being badly computed by both; and when combined with Algorithm 4, these figures become respectively 19.9% and 19.3%, 936 pieces being badly computed by both.

Therefore, the use of different feature sets seems to imply a significant amount of specific failures.

A more detailed analysis is shown on Figure 4.2 which plots the number of failures (normalized by the number of pieces in each genre) specific to each set, when combined with Algorithm 4, versus musical genres.[4] The first two bars in this figure should be understood as follows: "around 7% of errors made by Algorithm 4 on the *Cha Cha* data set when using feature set 6 were not made when using feature set 14 and, inversely, around 4% of errors made when using feature set 14 were not made when using feature set 6." In this figure, we can see for instance that all the errors made

---

[4]In this figure, we use the following abbreviations for musical genres: "Ch" stands for *ChaCha*, "Cl" for *Classical*, "El" for *Electronic*, "Ja" for *Jazz*, "Ji" for *Jive*, "Qu" for *Quickstep*, "La" for *Latin*, "Lo" for *Loops*, "Mi" for *Miscellaneous*, "Po" for *Pop*, "Ra" for *Rap/Hip-Hop/Trip-Hop*, "Re" for *Reggae*, "Ro" for *Rock*, "Ru" for *Rumba*, "Sa" for *Samba*, "So" for *Soul*, "Ta" for *Tango*, "VW" for *Viennese Waltz* and finally "Wa" for *Waltz*.

by set 14 on the *Jive* and *Pop* data sets are also made by set 6 but not the other way around, a significant number of errors that are made by set 6 are not made by set 14. The inverse is true on the *Latin* data set. We can also see that both sets make around the same number of specific errors on e.g. the *Classical, Loops, Rock, Soul* and *Waltz* data sets.

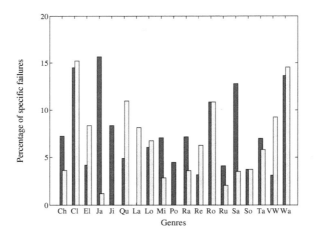

Figure 4.2:   Specific failures induced on pieces of different musical genres by two feature sets when combined with Algorithm 4: set 6 (in dark bars, the magnitude-normalized first-order differential of energy in the 36 ERB subbands) and set 14 (in light bars, the selection of features advocated on page 137, excluding magnitude-normalized first-order differentials of the energy in the 36 ERB bands).

As reported in Table 4.2 and in Appendix A, feature sets 6 (made up *only* of energy features) and 14 (free of such features but one, the first-order differential of MFCC1) are both relatively good sets; additionally, they are *complementary* as they make a significant amount of their errors on different pieces. This observation is true when considering different tempo induction algorithms.

These two feature sets seem to represent *different aspects of rhythm* and should therefore be considered jointly. One would expect that the combination of these two feature sets into a single set (i.e. feature set 13) would result in a significant performance increase. For instance, one may hope that the errors made by a *single* set would not occur when considering both sets together, hopefully, only the errors common to both sets would remain. If this were true, feature set 13 would reach an accuracy (on the whole data set and with respect to accuracy 2) of around 73.6% when combined with Algorithm 1. The accuracy actually reached is 63.8%, which is not significantly better than that obtained with the best individual set (set 14, 63.2%), the accuracy loss with respect to what would ideally be reached is around 10 percent points, which is considerable.[5] Similarly, when combined with Algorithm 4, the loss is around 6 percent points, when combined with Algorithm 12, the loss is around 4 percent points and in both cases the combination of sets does not perform significantly better than the best of the two sets (set 6 in both cases).

Let us consider the errors made by feature set 13. When associated with Algorithm 1, 69.5% of the errors made by the combination of sets 6 and 14 (i.e. set 13) are errors common to both sets. On the other hand, 20.2% are errors specific to set 6 when considered alone, 8.1% are errors specific to set 14 and 2.1% are errors that none of these sets makes when considered alone. When associated with Algorithm 4, these figures become respectively 80.4%, 13.4%, 5.6% and 0.5%. Finally, when associated with Algorithm 12, these figures become respectively 72.4%, 16.5%, 7.5% and 3.6%.

A conclusion from this analysis is that the strategies for combining features considered in this section do not yet take full advantage of large sets of good and complementary features. An important issue still resides in this aspect of tempo induction algorithms.

Joint estimation of several metrical levels   As can be observed in Table 4.2 (underlined results), the musical parsing of periodicity functions proposed by Dixon et al. (2003) is a better strategy than all the other combining and parsing strategies.

This is a strong argument in favor of the consideration of influential schemes between metrical levels in the implementation of tempo induction algorithms, and this confirms analyses of the ISMIR 2004 contest results (see on page 85).

---

[5]remark additionally that this is the *best* accuracy obtained by Algorithm 1.

Feature sets	Algorithms											
	1	2	3	4	5	6	7	8	9	12	13	14
	Integrating features			Integration rationale								
				Integrating periodicity functions								
	Sum			Sum			Product			Musical Parsing		
(Periodicity function)	ACF	comb	FFT	ACF	comb	FFT	ACF	comb	FFT	ACF	comb	FFT
set 1	68.1	74.2	68.3	68.1	74.2	68.3	68.1	74.2	68.3	<u>81</u>	77.5	**80.1**
set 2	64.2	73.2	67.3	64.2	73.2	67.3	64.2	73.2	67.3	<u>77.1</u>	76.7	74.6
set 3	69.7	80.6	69.5	73.2	83.8	70.5	**72.6**	83.4	70.1	**88.5**	**<u>90</u>**	69.9
set 4	67.9	77.7	64.6	71	**86.1**	53.8	68.1	82.8	63	<u>87.9</u>	87.5	52.1
set 5	61.1	78.1	65.2	67.7	81.4	61.8	64.8	76.3	63.8	<u>84</u>	83	54
set 6	72.2	77.1	67.9	68.9	72.8	**70.8**	71	74	68.5	<u>85.9</u>	82.6	51.1
set 7	70.7	79.1	69.5	70.5	78.5	70.5	71	78.1	68.7	85.7	<u>85.7</u>	75.1
set 8	61.5	72.4	63.2	68.1	73	68.3	68.3	73.2	65.6	<u>82.4</u>	76.5	36.8
set 9	53	68.3	51.5	60.1	58.3	55	4.1	35.8	56	<u>72.2</u>	54.2	24.1
set 10	60.3	75.3	65	64	74.6	64.4	61.8	72.6	63.6	<u>78.7</u>	73.6	37.6
set 11	**73**	**82.2**	**71.4**	**73.4**	84.7	69.3	70.8	**83.6**	71.2	<u>87.1</u>	86.9	70.8
set 12	64.2	78.1	64.8	65	76.9	65	65.4	77.3	65.8	82.6	<u>83.2</u>	69.1
set 13	71.8	78.7	68.5	71.2	73.2	68.1	69.5	74	**71.4**	<u>86.5</u>	84.9	49.1
set 14	65.8	80	66.5	71.4	76.5	68.1	71.2	74.4	70.1	<u>84.9</u>	83.6	40.7

Table 4.2: Accuracy 2, in %, of diverse tempo induction algorithms implementing different choices regarding feature sets, integration rationales (integrating features or periodicity functions), integration operations (sum, product or musical parsing) and periodicity functions (ACF, comb filterbank or Fourier transform). The evaluation is done over Alonso data. Bold fonts are used to highlight the best feature set given an algorithm (i.e. best line given a column), underlining is used to indicate the best algorithm given a feature set (i.e. best column given a line). Performances of Algorithm 10 and Algorithm 11 are not provided here for being clearly below those of the other algorithms.

4.3 Comparison with state-of-the-art tempo induction systems

In this section, we compare three of the best algorithms proposed above with state-of-the-art algorithms that took part in the ISMIR 2004 tempo induction contest.

In this comparison, we use the whole data set (i.e. 3223 pieces). Recall that 2734 pieces of this data set have been used for the contest (*Loops* and *Ballroom* data sets: 2036 + 698 pieces). It is therefore possible to extend the comparison presented here to the remaining algorithms that took part in the contest. As precised in (Gouyon et al., 2006), the 698 Ballroom pieces are publicly available and the 2036 Loops can be easily obtained, opening the way to researchers to compare their algorithms to those presented here.

Note that the (slight) differences between the performances reported here and those published in (Gouyon et al., 2006) are due to a change in the accuracy metrics (namely a slight increase of the precision windows).

From the pool of contest algorithms, we focus on the following one:

Klapuri: This algorithm, published in (Klapuri et al., 2005), has been described earlier in this book on page 63. We include it in this comparison as it won the ISMIR 2004 tempo induction contest.

Scheirer: We include Scheirer's algorithm (1998) in this comparison as it is a very influential algorithm, has represented for a long time the state-of-the-art in tempo induction and is open source and available on the web.[6] It has been described on page 64. Recall that a back-end has been added to the original code in order to output a single tempo scalar instead of a series of beats.

DixonACF: The algorithm by (Dixon et al., 2003) has been described on page 62. The reasons to include it in this comparison are its very good performance in the contest and the fact that Algorithm 12, also used in this comparison, implements the same combining and parsing strategy.

---

[6]http://sound.media.mit.edu/~eds/beat/tapping.tar.gz

We compare these algorithms to Algorithm 12 associated with three different feature sets: the 59 features selected on page 137 (set 13), the magnitude-normalized first-order differential of energy in Dixon's 8 subbands (set 3) and the energy in the 36 ERB subbands (set 4).

Algorithms	**All**		**Ballroom**		**Loops**		**Alonso**	
	acc1	acc2	acc1	acc2	acc1	acc2	acc1	acc2
Klapuri	68.3	85.9	63.9	91.8	71.1	82.3	63	92.6
Scheirer	38.9	69.7	52.3	75.9	33.2	66.2	43.6	75.7
DixonACF	42.6	85	43.5	87.8	43.3	83.7	38.4	86.7
Algo. 12 & set 13	46.3	84.7	57.9	91.3	44.2	82.1	35.4	86.5
Algo. 12 & set 3	54.7	84.9	63.5	88.1	53.9	82.9	45.8	88.5
Algo. 12 & set 4	63.8	83.4	61.6	83.2	67.3	83.4	51.9	87.9

Table 4.3: Comparison with state-of-the-art tempo induction systems.

Accuracies of the diverse algorithms are reported in Table 4.3. We can observe that our algorithms perform better than Scheirer on all data sets and with respect to both accuracy measures. They perform only slightly worse than Klapuri on overall, with respect to accuracy 2. On the other hand, they perform significantly worse with respect to accuracy 1.

On overall, our algorithms perform similarly to DixonACF when compared with accuracy 2 and perform better when compared with accuracy 1. This is interesting as it highlights the importance of the input features (only differences with DixonACF). Interestingly, even if it performs similarly with respect to accuracy 2, the association of Algorithm 12 with feature set 3 is significantly better than DixonACF with respect to accuracy 1. This is interesting as they are basically the same algorithm and use almost the same features, the difference lies (for the former) in the normalization of the feature differentials by the magnitude.

4.4 Conclusions

This chapter addressed the task of automatic tempo induction. We evaluated several tempo induction algorithms implementing different strategies regarding some open issues among those listed on page 82, namely the choice of periodicity function and the combination and parsing of multiple information sources. These algorithms' inputs are diverse low-level feature sets, ranked in Chapter 3 with respect to their representativeness of beats. Some algorithms show performances that are comparable to the state-of-the-art. We also demonstrated that a special effort should be dedicated to the choice of algorithm input features.

In accordance with findings of the previous chapter, energy in several frequency bands are better features than the energy computed on the whole frequency range, energy in low and high frequency bands are better features than energy in mid-frequency bands, spectral features and MFCCs are also relatively good features for tempo induction, first-order differentials of MFCCs are better than magnitude-normalized first-order differentials (indeed, the magnitude-normalized first-order differential of MFCCs is the worst feature set is the same for all tempo induction algorithms tested here), and the feature set selected on page 137 also provides relatively good accuracy figures.

Among many possible strategies for combining and parsing multiple information sources, the best strategy seemed to be that proposed by Dixon et al. (2003). This is a strong argument in favor of considering constraints posed by the metrical hierarchy in the design of tempo induction algorithms (i.e., estimating several metrical levels jointly instead of a single one). Results show that the Fourier transform is a worse rhythm periodicity function (at least in our use of it) than the ACF and comb filterbanks. Comb filterbanks seem to perform slightly better than the ACF, but no clear conclusion could be reached. More research is still required in the design of mathematical transformations suited to the computation of rhythmic periodicity functions. Integrating periodicity functions seems to yield slightly better results than the integration of features. Nevertheless, the slight accuracy gain may not be worth the increase in computation load.

As detailed in this chapter, feature set rankings (from Chapter 3) are not verified for all algorithms. That is, when used as input to some algorithms, feature sets that are highly representative of beats may provide worse results than less representative feature sets. In our opinion, this indicates that the tempo induction algorithms used here are still not taking full advantage of the explanatory power of feature sets and that more research is needed in the design of better algorithms. We have seen for instance in Paragraph 4.2.3.3 that the combination of good and *complementary* feature sets does not necessarily yield significantly better results than their parts. More efforts should therefore be dedicated to improving strategies for combining multiple information sources. Indeed, it is possible that the combining and parsing strategies used here are sensitive to the number of features and impose a sort of "ceiling" in terms of number of feature they can deal with, hence preventing the use of high numbers of features. This would explain the fact that the ranking, according to Chapter 3, of feature sets 6 (36 features), 3 (8 features) and 7 (4 features) are not respected by many algorithms (see on page 152). This would also explain the fact that the "best" feature set (i.e. the most representative of beats, set 13, made up of 59 features), even if yielding results close to the best when combined with most algorithms, does not always outperform other feature sets.

Rhythm periodicity functions plot salience versus period (or frequency) via processes that emphasize self-similarity in a given signal. This is closely related to the process of template matching, where the template is defined by the data at hand. Viewing template matching as one among several approaches to pattern recognition (which associated recognition function can be e.g. the correlation, (Jain et al., 2000, Table 2)), a generalization leads to the potentially interesting consideration of other pattern recognition approaches in the task of seeking periodic behaviors in feature lists. Pattern recognition techniques would permit a more elegant way to integrate the reduction of the number of features in this task and may be more effective in dealing with large numbers of features than the simple techniques used in this chapter.

Chapter 5

Applications of rhythm periodicity functions for music content processing

The standardization of personal computers and worldwide low-latency networks, the extensive use of efficient search engines in everyday life, the continuously growing amount of multimedia information on the web, in broadcast data streams or in personal and professional databases and the rapid development of on-line music stores such as Apples iTunes has recently boosted developments in Music Information Retrieval (MIR) and music content processing.

MIR is a young and very active research area. This is clearly shown in the constantly growing number and subjects of articles published in the Proceedings of the annual International Conference on Music Information Retrieval (ISMIR, the first established international scientific forum for researchers involved in MIR) and also in related conferences and scientific journals such as ACM Multimedia, IEEE International Conference on Multimedia and Expo or Wedelmusic, to name a few. Research in MIR addresses the wealth of scenarios for interacting with music posed by the digital technologies in the last decades. Applications are manifold, consider for instance automated music analysis, personalized music recommendation, intelligent on-line music access, query-based retrieval (e.g. "by-humming," "by-example") and auto-matic play-list generation. For Aigrain (1999), in the near future, content-processing technologies will provide "new aspects of listening, interacting with music, finding and comparing music, performing it, editing it, exchanging music with others or sell-ing it, teaching it, analyzing it, and criticizing it." Along this line of thought, we understand by "processing music content" both the *exploitation* and *description* of music content. That is, as we argue in (Gouyon et al., in press), "processing" may mean describing a musical database; browsing, exploring or understanding pieces in such a database; comparing a number of pieces; retrieving any desired selection of pieces; introducing new pieces to a database; recommending music (Cano et al., 2005b); transforming pieces; etc. Among the vast number of disciplines and ap-proaches to MIR (an overview of which can be found in (Downie, 2003)), part of the research is dedicated to the extraction of musical descriptors from audio signals. Pro-cessing, in its diverse meanings (retrieval, transformation, etc.), is applied on these descriptors. Among these descriptors, rhythm descriptors are especially relevant. For instance, tempo plays an important role in automatic sequencing of musical pieces into playlists (for dancing at least) and rhythmic expressiveness transformations. In addition to tempo, rhythmic descriptors as the swing or the time signature determine partly musical genres; they are therefore of very first relevance in automatic genre classification, as in many other MIR applications.

This chapter illustrates the use of rhythm periodicity functions and descriptors de-rived from such functions in MIR scenarios: Section 5.1 illustrates the use of tempo, tatum, time signature and periodicity features in genre classification experiments. Section 5.2 illustrates the use of another descriptor, the swing, in rhythmic expres-siveness transformations.[1]

---

[1]Note that in this chapter we do not use the periodicity functions advocated in previous chapters. This is because the work reported in this chapter was previous to that reported in Chapters 3 and 4. However, the purpose of this chapter is to demonstrate the use of periodicity functions in general, not that of the best periodicity function.

5.1 Genre classification

Musical genre is a fundamental kind of metadata for browsing musical collections. Indeed, people often describe their musical tastes with respect to genre. Musical genre classification has received much attention from music record retailers and, recently, from audio and music researchers, especially in the MIR community (Tzanetakis and Cook, 2002). An important direction of research now relates to the definition of features of musical genres and their automatic extraction from various forms of musical data (audio, scores, MIDI, MP3, etc.). Even if there is still room for disagreement on explicit definitions of musical genres (Aucouturier and Pachet, 2003), there is a pervasive belief that this notion has something to do with fundamental musical dimensions such as melody, instrumentation, harmony and rhythm. Rhythmic descriptors are therefore very valuable candidates for musical metadata. For instance, Dixon et al. (2003) claim that very few periodicities (the tempo, the measure and optionally others as the dotted quarter-note) seem sufficient to classify 8 rhythmic classes relatively well. Tzanetakis and Cook (2002) and Pampalk et al. (2003) respectively report on genre classification experiments and definitions of similarity distances using signal descriptors embedding (somehow) rhythmic aspects. Also, Foote et al. (2002) claim that all aspects of rhythm are captured by a specific periodicity representation and that such representation is sufficient for the retrieval of similar pieces of audio. This conclusion is however based on the analysis of solely 15 musical excerpts (4 songs divided into several 10 s chunks).

In this chapter, we assess the relevance of a set of rhythmic descriptors in automatic musical genre classification experiments. The Ballroom data set introduced on Paragraph 2.3.2.2 provides the necessary ground-truth for our experiments.[2] We acknowledge that there actually exists no ground-truth with respect to genres (Aucouturier and Pachet 2003). However, some musical genres are rapidly recognizable by listeners, even with minimal musical training, and on the dance floor, dancers do recognize instantly what dancing step fits best to the music they hear. Dancing having much to do with rhythm, we believe that ballroom dance music provides a relatively solid basis for our

---

[2]Recall that this data consists of 8 classes of ballroom dance music, it is publicly available and the classification random-guess baseline accuracy is 15.9%.

experiments.

Sections 5.1.1, 5.1.2, 5.1.3 and 5.1.4 illustrate the use of, respectively, the tempo, the tatum, the time signature and periodicity features in genre classification experiments.

5.1.1 Tempo

5.1.1.1 Algorithm

Many algorithms exist to compute the tempo of audio signals, they are reviewed in Chapter 2. As claimed in Chapter 2, they all imply the computation of a periodicity function. We will focus here on the BeatRoot algorithm from Dixon (2001a), available as GPL code.[3] This algorithm is referred as DixonT and detailed on page 62.

5.1.1.2 Genre classification experiments

As reference, let us consider the *correct* tempo alone (i.e. measured *manually*). A 1-NN classifier using solely this descriptor classifies the 8 ballroom dance classes with an accuracy of 82.3%. A C4.5 decision tree achieves 78.6%. This last result was obtained with a special tweaking of the algorithm: forcing a relatively high number of instances per leaf,[4] which results in smaller trees, with fewer leaves and guarantees good generalization of the result. The number of leaves is 9. Each class corresponds to a leaf, except one (Rumba) which corresponds to two leaves. In sum, this technique highlights a clear ordering of classes with respect to tempi. Therefore, one can assume that, given a musical genre, the tempo of any instance is among a very limited set of possible tempi, namely, from slow to fast:[5]

---

[3]at http://www.ofai.at/~simon.dixon/beatroot/index.html. Different algorithms could be used for the task, as those detailed in other chapters of this document.

[4]20 instead of default value 2

[5]Note however that these high levels of accuracy may be due to the relatively small number of instances (698). The observation that tempo is genre-specific may not hold when considering very large number of musical pieces.

tempo< 91	⇒	Slow Waltz
91 <tempo< 96	⇒	Rumba
96 <tempo< 102	⇒	Samba
102 <tempo< 104	⇒	Rumba
104 <tempo< 124	⇒	Cha Cha
124 <tempo< 141	⇒	Tango
141 <tempo< 176	⇒	Jive
176 <tempo< 180	⇒	Viennese Waltz
tempo> 180	⇒	Quickstep

Using solely BeatRoot tempo, a 1-NN classifier yields 51.7% correct classification. With C4.5, an accuracy of 52.5% is achieved. As illustrated in Figure 2.12(a), the algorithm makes systematic errors by confusing metrical levels. Metrical level errors cause the loss of accuracy with respect with using the correct tempo. More importantly, this causes decision tree to have too many leaves (around 15). This means that the tempo axis is divided in small clusters which do not represent characteristic tempo zones.

5.1.2 Tatum

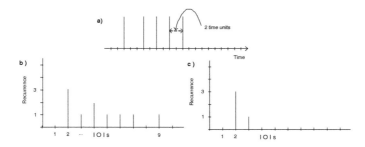

Figure 5.1: Onset sequence (a) — IOI histograms (b and c).

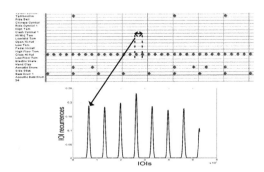

Figure 5.2: "Piano Roll" and IOI smoothed histogram of a MIDI drum track.

5.1.2.1 Algorithm

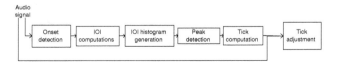

Figure 5.3: Tatum induction algorithm flow diagram.

Fundamental to the tatum induction algorithm is the computation of audio signal Inter-Onset Intervals (IOIs, subsequent to a detection of onsets). In accordance with the tatum definition (page 22), keeping the shortest IOI would not suffice to determine the tatum. Indeed, in e.g. syncopated musical excerpt, the tatum may not be explicit in the IOI list, but rather be implied by the relationships between those intervals (see Figure 5.1 for an illustration). Such cases are better handled when defining an IOI as the time difference between any two onsets (not necessarily successive) than between successive onsets. Here also, see Figure 5.1 for an illustration: in b) IOIs are computed taking into account all pairs of onsets, in c) IOIs are computed taking into account solely successive onsets, the tatum —of 1 time unit— is not explicit. Therefore, the algorithm is based on a measure of IOI recurrence. As there are integer timing ratios between metrical levels, histograms of IOIs should show peaks at approximately harmonic positions. If one extracts note-on timing data from quantized MIDI drum tracks, then the fact that the fastest pulse contributes to the raising of peaks in the histogram at the exact positions of all of its multiples can be clarified visually on Figure 5.2. Therefore, herein the tatum is defined as the gap of the IOI histogram harmonic series —one could make an analogy with the notion of fundamental frequency. The tatum induction algorithm is divided in the following steps.[6]

Onset detection The short-time energy is computed over non-overlapping signal frames (e.g. 11 ms). When the energy value is a certain percentage (e.g. 200%) higher than the energy average of a fixed number of previous frames (e.g. 8), an onset is detected. It is assumed that there are at least 60 ms in between two onsets. To each onset is associated a weight (i.e. a degree of confidence), corresponding to the number of after-onset successive frames whose energy is higher than the aforementioned averaged energy. The weight gives an indication whether the onset should be considered as an actual one or an artifact of the onset detection scheme, which can be useful for subsequent uses of the onset list. Optionally, a minimum number of onsets per second (e.g. 2.5) can be imposed to the algorithm. To reach this requisite, the aforementioned percentage is lowered step by step (step set to e.g. 10%).

IOI computations As mentioned earlier, we take into account the time differences between any two onsets. A weight is associated to each IOI, corresponding to the smallest weight among the two onsets used for the IOI computation.

IOI histogram generation In order to handle short-time timing deviations, the histogram is smoothed by convolution with a Gaussian function whose standard deviation was empirically adjusted.

---

[6]Part of the material in this section has been previously published in a conference article written with Perfecto Herrera and Pedro Cano (Gouyon et al., 2002).

Algorithm 1 TWM first error function.
___
1: deviation = 0
2: **for** each candidate pulse period **do**
3:    **for** each candidate pulse phase **do**
4:       **for** each IOI histogram peak **do**
5:          deviation ← deviation+distance of the peak to the closest beat
6:       **end for**
7:    **end for**
8: **end for**

IOI histogram peaks detection Peak positions and heights are detected in the histogram with a 5-point running window method. A local maximum is detected at index i when the corresponding value is higher than four others: at indexes $i - 2k$, $i - k$, $i + k$ and $i + 2k$ (where k is set to e.g. 4).

Tatum period computation The fundamental aspect of the tatum computation is the use of a particular pulse matching function (see page 45): the Two-Way Mismatch error function (TWM, (Maher and Beauchamp, 1993)). According to the previous definition of the tatum, we seek the IOI that best predicts the harmonicity of the IOI histogram. The basic procedure is to consider many possible tatum period candidates (all with phase zero) and generate corresponding pulses and measure their matching with IOI histogram peaks. Candidate periods range e.g. from 80 ms to 700 ms (750 BPM to 85 BPM). For each tatum candidate, a corresponding pulse is generated and two error functions are computed. The first one illustrates how well the IOI histogram peaks explain the beats (resembling the "positive evidence" mentioned on page 40): a global deviation is computed as depicted in the algorithm on this page. The second one illustrates how well the beats explain the IOI histogram peaks (resembling the "negative evidence" mentioned on page 40); it is computed as the number of unmatched beats. The TWM error function is a linear combination of these two functions (e.g. with equal weighting factors). The tatum period is set to the period of the pulse corresponding to the TWM error function global minimum.

Tatum period adjustment and phase computation   The tatum period is re-
fined and its phase computed by achieving a matching between several pulses and the
signal onsets. As in the previous step, the TWM error function is computed. Differences are that the matching is done on the onsets (and not the IOI histogram peaks),
that few period candidates are considered around the first tatum period estimation
(e.g. period$\pm$15 ms, by 1 ms steps) and that we do seek the best phase.

The reason for achieving the tatum period adjustment is that the first tatum
period estimation is not very accurate because of the histogram smoothing. The
smoothing permits to agglomerate IOIs that should be considered jointly, even if
not exactly equal; it necessarily entails a trade-off between precision and amount of
IOI agglomeration. Note that the accuracy of the tatum period computation is an
important issue, even a very small error in this value does propagate in an additive
manner in the prediction of future tatum beats.

5.1.2.2   Algorithm evaluation

The algorithm has been evaluated on drum tracks,[7] both artificially generated (1000
instances) and real ones (57 instances). The algorithm for generating artificial drum
tracks is roughly the following, a primary pulse is first defined at each integer multiple
position of which a percussive sample is added in an empty audio signal. Samples
are randomly chosen from a percussive sound database. Then, a tatum is defined as
an integer divisor of the previous pulse, at each position in the tatum track, either
a sample or silence is randomly assigned. To account for more realistic features,
deviations of 1 to 10 ms from the exact tatum track positions are allowed and white
noise with a SNR of 30 dB is subsequently added to the signal. The ground-truth
for the tatum period is an input parameter for this algorithm. On the other hand,
the tatum period of 57 real drum tracks have been manually annotated. Using the
accuracy measures detailed on page 70, performance are summarized in Table 5.1.

---

[7]Audio signals of restricted polyphonic complexity, containing specific sets of percussive timbres
as acoustic bass drums, snare drums, hi-hats, toms and cymbals.

Drum tracks	Accuracy 1	Accuracy 2
Artificial	77.3	88.4
Real	86	93

Table 5.1: Performances (in %) of the tatum-finding algorithm on drum tracks

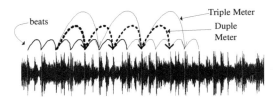

Figure 5.4: Illustration of the two possibilities for beat grouping: duple or triple?

5.1.2.3   Genre classification experiments

Using the tatum as single input feature of a 1-NN classifier yields a classification accuracy of 50.4%, using a C4.5 tree the accuracy is 51.8% (15 leaves), both accuracies are computed as 10-fold cross-validations. These accuracies are comparable to those obtained when using the tempo (see on page 166).

5.1.3   Time signature

In this section we address the problem of classifying polyphonic musical audio signals by their meter: the number of beats between regularly recurring accents (or downbeats). The problem is simplified to a 'duple' vs 'triple' decision (i.e. groupings of two beats vs groupings of three beats), see Figure 5.4 for an illustration).[8]

---

[8]Part of the material in this section has been previously published in a conference article written with Perfecto Herrera (Gouyon and Herrera, 2003b).

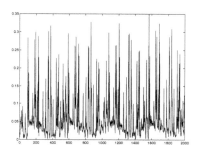

Figure 5.5: Evolution of the energy over the frames of 20 seconds of "A lo Cubano" (*Orishas*, Cuban Hip-Hop)

5.1.3.1 Algorithm

The approach followed in this algorithm aims to test the hypothesis that acoustic evidences for downbeats can be measured on signal low-level features; focusing especially on their temporal recurrences. The algorithm is based on the computation of inter-beat segment features, beat indexes being extracted in a semi-automatic manner to provide reliable input to the problem of interest here.

Frame feature computation We set a frame size of 20 ms, and a hop size of 10 ms. On each signal frame, the following low-level features are computed:[9]

1. f_1: Energy

2. f_2: Spectral flatness, i.e. ratio geometric mean/arithmetic mean (for this feature, frames are multiplied by a Hamming window before DFT computation)

3. f_3: Energy in upper-half of first Bark band (approximately 50-100 Hz)

Tactus induction As previously mentioned, beat indexes are extracted in a semi-automatic manner to provide reliable input to the problem of interest here. (Chapter 2 provides a review of algorithms that could be used for the task.)

---

[9]More details on the determination of the best features (and the inter-beat segment descriptors) by diverse feature selection experiments can be found in (Gouyon and Herrera, 2003b).

Region definition Beats are matched with frame indexes. For each beat, three regions of interest are defined:

1. $R0$: The whole segment between the beat and the next one (the inter-beat segment), recentered around the beat

2. $R1$: The 120 ms region surrounding the beat

3. $R2$: The rest of the inter-beat segment, i.e. $\left(R0 \cap \overline{R1}\right)$

Inter-beat segment descriptor computation Four descriptors are defined as the standard deviation of f_1 over $R0$, the average of f_2 and f_3 over $R1$, and the temporal centroid over $R0$ (this descriptor does not entail frame feature computation). Values of the descriptors are normalized (mean is subtracted and they are divided by the standard deviation). Each musical excerpt is then represented by 4 temporal sequences, whose lengths correspond to the number of beats of this excerpt. Each sequence is the evolution of a specific descriptor over the different inter-beat segments, see Figure 5.6 for an illustration.

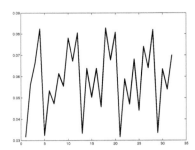

Figure 5.6: Evolution of the energy standard deviation over $R0$s (same song, same temporal scale on the X-axis as on Figure 5.5, but measured in beat indexes).

Periodicity detection The (normalized) autocorrelation $r(\tau)$ is computed for each sequence as follows. Let x be the subsequence corresponding to beat indexes 0 to I, and y the subsequence corresponding to beat indexes τ to $(\tau + I)$.

$$r\left(\tau\right) = \frac{\sum_{i=0}^{I} x_i y_i}{\sqrt{\sum_{i=0}^{I}\left(x_i\right)^2}\sqrt{\sum_{i=0}^{I}\left(y_i\right)^2}} \qquad \forall \tau \in \{0 \cdots U\}$$

where U is the upper limit for the lag τ (e.g. 8 beats), and I the integration time (e.g. 10 beats). High peaks in a descriptor autocorrelation function indicate lags for whose this descriptor reveals recurrences along the sequence.

Computation of decisional features We specify the criterion M for making decisions regarding the duple or triple nature of excerpts:

$$M = \left(\frac{\left(r\left(2\right) + r\left(4\right) + r\left(8\right)\right)}{3}\right) - \left(\frac{\left(r\left(3\right) + r\left(6\right)\right)}{2}\right)$$

M is a real number; the farther from zero in the positive values, the more it is representative of duple time signatures; the farther from zero in the negative values, the more it represents triple time signatures. There is one value of M for each descriptor. Henceforth, the relevant features for the duple/triple decision are the values of M corresponding to each descriptor. In the example illustrated in Figures 5.5 and 5.6, relative to a single feature (the evolution of the energy standard deviation over $R0$s), $M = 0.6313$, the time signature is effectively duple.

Classification Excerpts are represented by 4 features: the criteria M relative to the 4 inter-beat segment descriptors. The decision regarding the time signature of a test excerpt shall be taken according to the set of values for these features. Deriving a class membership from a set of descriptor values can be achieved by several pattern recognition techniques. For instance, Discriminant Analysis (DA) derives regions of class memberships in the 4-dimensional feature space from labeled data —in our case, by the very definition of M, the region boundaries are around zero. This technique gave us fairly good results. Even a simple rule, relative to a single feature, seems to give error rates relatively acceptable. Namely, "For a given excerpt, if $M_{temporal\ centroid(R0)} > -0.108046$, then this excerpt has a duple time signature, otherwise its time signature is triple."

5.1.3.2   Algorithm evaluation

The CUIDADO data set (see Section 3.2) was used for evaluating this algorithm. Recall that it is made up of 70 sounds, 34 of which are triple and 36 duple. We have tested different approaches to classification ranging from non-parametric models (kernel density estimation) to parametric ones (discriminant analysis), and including rule induction, neural networks, 1-Nearest Neighbor (1-NN), or Support Vector Machines (SVMs). Results are obtained by ten-fold cross-validation. A discriminant analysis with the four features introduced above yields a 5.2% error rate. With 6 different classifiers (Naïve Bayes, kernel density, 1-NN, SVM, C4.5, PART)[10] and a single feature, the feature M computed from the temporal centroid values over $R0$s, error rates were found to lie around 10%.

5.1.4   Periodicity features

5.1.4.1   Descriptors

We consider a total of 69 descriptors.[11]

Periodicity Histogram descriptors   Eleven descriptors are based on a first representation of signal periodicities, the "periodicity histogram" (PH) (Pampalk et al., 2003). This representation (see Figure 5.7), loosely inspired by (Tzanetakis and Cook, 2002), is the collection in a histogram of the saliences of different pulses (from 40 BPM to 240 BPM) in successive chunks of signal (12 s long, with overlap). In each chunk of signal, periodicities are computed via a comb filterbank (Scheirer, 1998). Among relevant differences with previous works stands the fact that the audio data is first preprocessed by a psychoacoustic model, removing information in the audio signal which is not critical to our hearing sensation while retaining the important parts.

---

[10]Experiments have been done using the commercial software Systat (http://www.systat.com/) and the open-source software Weka.

[11]Part of the material in this section has been previously published in a conference article written with Simon Dixon, Elias Pampalk and Gerhard Widmer (Gouyon et al., 2004a).

Also, periodicity magnitudes are weighted with respect to their periods, emphasis being given to tempi around 120 BPM, the "preferred tempo" region.[12]

Descriptors are the following:

- The most salient periodicity: highest peak in the PH.

- The distinctiveness of the most salient periodicity. It is measured as the ratio between the highest peak and the second highest peak.

- The periodicity power. This is the sum of the energy in the PH.

- The periodic energy in the first three Bark bands. This is the same as the previous, but considering solely the energy in the 3 lowest frequency bands defined by the Bark scale, below 300 Hz.

- The PH centroid, defined as the tempo for which half of the PH energy is contained in lower tempi.

- Three measures of the percussiveness. The percussiveness is computed as the central tendency of the energy in diverse frequency bands, defined by the Bark scale, of the half-wave rectified, first-order difference filtered, waveform. We use three variations of this descriptor where the central tendency of the energy is computed in different ways: $mean(x)$, $mean(x > mean(x))$ and $median(x > median(x))$.

- Three measures of the percussiveness in low frequencies. This is similar as above but using only the energy in the 3 lowest Bark bands.

Inter-Onset Interval Histogram descriptors Another pool of 58 descriptors is made up of quantities computed from a second representation of the signal periodicities, the Inter-Onset Interval Histogram (IOIH) introduced by Gouyon et al. (2002) and described in Section 5.1.2. Time intervals (in seconds) are drawn on the X-axis while (normalized) recurrences are drawn on the Y-axis (see Figure 5.8).

---

[12]See (Pampalk et al., 2003) for details.

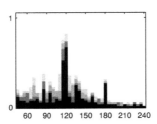

Figure 5.7: Periodicity histogram of a Jive excerpt. The tempo is 176 BPM. Gray shadings tell us the number of analysis chunks for which a certain energy was exceeded. Note the effect of "preferred tempo" weighting.

The IOIH is in many ways similar to the IOI clusters obtained in (Dixon et al., 2003). We therefore compute descriptors directly inspired by those detailed in (Dixon et al., 2003): selected prominent periods in the IOIH, together with their saliences.

- The saliences of 10 periodicities whose periods are the 10 first integer multiples of the tatum. Note that solely the period *salience* is kept, not the period value. Therefore, those descriptors are independent of the tempo.

Then, inspired by the analogy between the IOIH and a spectral representation, we define 48 other descriptors as common "spectral" descriptors (distribution statistics and MFCCs), but computed on the IOIH, not on a spectrum. In the following $\{x_i\}_{i=1...N}$ are the IOIH samples.

- The mean of the IOIH magnitude distribution

$$\frac{\sum_{i=1}^{N} x_i}{N}$$

- The geometric mean of the IOIH magnitude distribution

$$(\prod_i x_i)^{1/N}$$

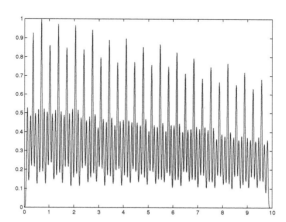

Figure 5.8: IOI histogram of the same Jive excerpt as Figure 5.7. Recurrence vs time interval. The tempo is 176 BPM (around 350 ms), which corresponds to the third peak (not to the highest one). The second highest peak is the measure (44 MPM, around 1.4s).

- The IOIH total energy

$$\sum_{i=1}^{N} x_i^2$$

- The IOIH centroid

$$\frac{\sum_{i=1}^{N} i x_i}{\sum_{i=1}^{N} x_i}$$

- The IOIH flatness

$$\ln\left(gmean\right) - \ln\left(mean\right)$$

- The kurtosis of the IOIH magnitude distribution. It measures how outlier-prone a distribution is, i.e. its degree of peakedness.[13]

$$\frac{\mu_4}{\mu_2^2} - 3$$

---

[13]http://mathworld.wolfram.com/topics/Moments.html

where μ_2 and μ_4 are respectively the second and fourth central moments of the IOIH magnitude distribution.

- The IOIH "high-frequency content"

$$\sum_{i=1}^{N} i x_i^2$$

- The skewness of the IOIH magnitude distribution. This is the degree of asymmetry of a distribution. A distribution spread out more to the left than to the right of the mean has a negative skewness. Perfect symmetry (e.g. a Gaussian distribution) results in a null skewness.

$$\frac{\mu_3}{\mu_2^{3/2}}$$

where μ_3 is the third central moment of the IOIH magnitude distribution.

- The first 40 coefficients of an analog to the Mel-Frequency Cepstral Coefficients (MFCCs). MFCCs are widespread descriptors in speech research. The Cepstral representation has been shown to be of prime importance in this field, partly because of its ability to nicely separate the representation of voice excitation (the higher coefficients) from the subsequent filtering performed by the vocal tract (the lower coefficients).[14] Roughly, lower coefficients represent the spectral envelope (i.e. the formants) while higher ones represent finer details of the spectrum. One way of computing the Mel-Frequency Cepstral representation of a time signal is detailed on page 108. In our case, we follow the same steps, but starting from the IOIH, not the magnitude spectrum. Note also that the number of coefficients is different (40 instead of 13).

5.1.4.2   Genre classification experiments

All classification accuracies reported below are computed as 10-fold cross-validations.

---

[14]http://mi.eng.cam.ac.uk/~ajr/SA95/

PH descriptors With the 11 PH descriptors, a 1-NN classifier yields 52.8% correct classification. The classification rate can be kept around the same value (slightly higher, 56.7%) when discarding 6 descriptors, and keeping solely:

- The most salient periodicity

- The distinctiveness of the most salient periodicity

- The periodicity power

- The PH centroid

- The first measure of the percussiveness in low frequencies

Periodicity saliences We refer here to the magnitudes of the IOIH peaks whose periods are the ten first integer multiples of the tatum.

When using the 10 IOIH peak amplitudes, keeping therefore solely descriptors that are *independent of the tempo*, we reach 51.2% of correct classification.

Other IOIH descriptors Let us consider the first 8 distribution statistics (i.e. not the MFCC-like). Using them all yields 46.1% classification accuracy with a 1-NN classifier. Selecting solely 3 yields a slight improvement: 48.7%. "Winning" descriptors are:

- The kurtosis

- The skewness

- The high-frequency content

Let us now consider the MFCC-like descriptors. Also with 1-NN classification, the whole pool (i.e. 40 descriptors) yields 79.6% accuracy. A very similar classification accuracy can be reached (79%) when selecting the following 15 coefficients: MFCC1, MFCC2, MFCC3, MFCC6, MFCC7, MFCC8, MFCC10, MFCC11, MFCC15, MFCC16, MFCC19, MFCC24, MFCC25, MFCC26 and MFCC28.

Understanding IOIH MFCC-like descriptors When dealing with speech signals, it has been shown that most of the relevant information occurs near the origin of the cepstral representation and in a few peaks higher up the cepstrum,these peaks corresponding to multiples of the pitch. Hence the focusing on the first MFCCs (less than 20), providing a compact representation of the spectral envelope while discarding the fine detail pitch information. This is especially true in speech recognition tasks where researchers precisely seek pitch-independent descriptors.

When dealing with music signals, and when replacing the Fourier transform by an ad-hoc transformation (the IOI histogram), it is less clear that higher coefficients should be discarded. In our case, higher coefficients provide a representation of finer detail of the IOIH peaks, that is, a closer look at the harmonic nature of this periodicity representation, its "pitch."[15] Therefore, higher coefficients seem to be somehow related to the pace of the piece at hand. On the other hand, lower coefficients represent the global envelope of the IOIH, which would be the "spectral envelope" of a proper spectrum. They seem to represent in some way the global structure of the IOIH.[16] In our understanding, they encode some aspects of the metrical hierarchy. Independently of the tempo.

5.1.5 Using expert classifiers

Section 5.1.1 shows that the correct tempo is a very relevant feature for genre classification and that given a musical genre, the tempo of any instance is among a very limited set of possible tempi. Moelants (2003) shows on a large amount of data (more than 90000 instances) that different dance music styles ("trance, Afro-American, house and fast") show clearly different tempo distributions, centered around different "typical" tempi.

However, the metrical level errors that are typical of tempo induction algorithms produce a dramatic decrease in classification accuracy.

---

[15]Note that the tatum is precisely computed as the "gap of the IOIH harmonic series" (Gouyon et al., 2002).

[16]For instance, excerpts whose periodicities have very similar saliences, as e.g. many Cha Cha, have a flat envelope.

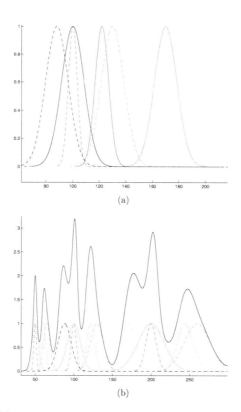

(a)

(b)

Figure 5.9: 5.9(a): Tempo probability functions of 8 dance music styles. X-axis in BPM. 5.9(b): Tempo probability adapted to typical metrical level errors, the solid black line is the sum of all probability functions and represents overall class overlaps.

In this section, we consider typical tempo induction errors in a domain-specific learning methodology, where the computed tempo is used to select an expert classifier which has been specialized on its own tempo range.[17] This enables the eight-class learning task to be reduced to a set of two- and three-class learning tasks. In this framework, a classifier focuses first on the tempo and then uses complementary features, as those detailed in Section 5.1.4, to make decisions in possibly ambiguous situations (i.e. tempo overlaps).

For instance, let us consider the tempo ranges given in Table 5.2 and illustrated in Figure 5.9(a). We define a Gaussian tempo probability function for each class. The Gaussian standard deviations are defined so that the probabilities at the limits specified in Table 5.2 are half the value of the corresponding probability maximum. Put together, these probabilities may overlap in certain tempo regions (e.g. Samba and Rumba, see dashed-blue and solid-black lines around 100 BPM in Figure 5.9(a)).

We can adapt the tempo probability function of each genre accordingly to typical tempo errors concatenating several Gaussians whose means are correct tempi and relevant multiples. See an illustration on Figure 5.9(b).

Genre	Tempo range
Cha Cha	$116 - 128$
Jive	$160 - 180$
Quickstep	$198 - 210$
Rumba	$90 - 110$
Samba	$96 - 104$
Slow Waltz	$78 - 98$
Tango	$120 - 140$
Viennese Waltz	$170 - 190$

Table 5.2: Dance music tempo ranges, in BPM.

---

[17]Part of the material in this section has been previously published in a conference article written with Simon Dixon (Gouyon and Dixon, 2004).

5.1.5.1 Algorithm

Observing the probability functions in Figure 5.9(b), one can see that each tempo value corresponds usually to two different potential classes, at the exception of three specific tempo regions in which three classes overlap. These are 95 to 105 BPM and 193 to 209 BPM for Quickstep, Rumba and Samba, and 117 to 127 BPM for Cha Cha, Tango and Viennese Waltz. Therefore, we propose to build 30 different classifiers:

- $\sum_{n=1}^{8-1} n = 28$ two-class classifiers, $\{K_1 \ldots K_{28}\}$, each expert in a specific pairwise classification task.

- 2 three-class classifiers, K_{29} and K_{30}, each expert in a three-class specific task

When presented with unknown instances, the knowledge available to the system is this set of 30 expert classifiers and the tempo probability functions for all possible classes. Therefore, the overall classification process is given in the algorithm on the current page.

Algorithm 2 Overall classification process

1: Compute tempo T of the instance to classify
2: Find the classifier K_i whose tempo range includes T
3: Perform classification with K_i

Descriptors In addition to the BeatRoot tempo, we consider the 69 periodicity features computed on two periodicity functions: the Periodicity Histogram and the Inter-Onset Interval Histogram detailed in Section 5.1.4.

For each of the 30 classification tasks, we discarded the use of the tempo and we evaluated the relevances of the remaining low-level descriptors on an individual basis (i.e. Ranker search method associated to ReliefF attribute evaluator),[18] and selected the 10 most relevant features. That is, the 30 classifiers all use 10 low-level features, that may be different in each case.

---

[18]Experiments have been conducted with Weka (Witten and Frank, 2000).

5.1.5.2 Genre classification experiments

For classification, we use Support Vector Machines as it is commonly suggested for problems with few classes (especially two-class problems). All percentages result from 10-fold cross-validation procedures.

The majority of the 28 pairwise classifier accuracies, all using 10 descriptors, are above 90%. The worst classifier is that between Slow Waltz and Viennese Waltz (81.8% accuracy, baseline 63%). The best is that between Quickstep and Viennese Waltz (100% accuracy, baseline 55.7%). Regarding the three-class classifiers, also using 10 descriptors, K_{29} (Quickstep vs Rumba vs Samba) has 84.1% accuracy (baseline 36.6%) and K_{30} (Cha Cha vs Tango vs Viennese Waltz) 91.9% accuracy (baseline 42.3%).

To measure the overall accuracy of the 30 classifiers, let us compute a weighted average of their individual accuracy. The weights are proportional to the number of times a classifier is actually required (given the tempo estimations of the 698 excerpts). This yields 89.4% accuracy.

Let us now evaluate the whole classification process. Recall that the process involves two steps, it suffers from tempo estimation errors in addition to misclassifications. In 24.3% of the cases (i.e. 170 excerpts) the tempo estimation step assigns excerpts to pairwise (or three-class) classifiers that do *not* account for its true class. There is no way to recover from these errors, whatever the subsequent classification, the excerpt will be assigned to an incorrect class.

The overall accuracy of the system is therefore the multiplication of both step accuracies, i.e. 0.894×0.757=67.6%.

One might wonder whether considering metrical level errors in the design of the tempo probabilities (i.e. using tempo probabilities as defined in Figure 5.9(b) instead of Figure 5.9(a)) actually results in any improvement. Recall that considering simple multiples of the correct tempo as errors BeatRoot tempo induction algorithm has an accuracy of around 50%. The resulting overall accuracy of the method presented here would therefore be around 0.894×0.5=44.7%. The improvement is over 20%.

However, we noted that tempo induction is especially bad for Slow Waltz excerpts, yielding around 75% to be assigned to wrong classifiers. This is because onset detection, in the tempo induction algorithm, is designed for percussive onsets, which are often lacking from waltzes. Removing the Slow Waltz excerpts for the database, 587 remain, and the number of excerpts that are assigned to irrelevant classifiers falls to 13.9%. The overall accuracy rises now to 76.5%.

In conclusion, reducing the problem from an eight-class learning task to several two- or three-class learning tasks is only pertinent when using an *extremely* reliable tempo estimation algorithm. To illustrate this, let us consider using the correct tempo (assigned manually) instead of BeatRoot tempo (computed automatically). There, the classification accuracy rises to 82.1%. This corresponds to two factors: misclassifications of the expert classifiers (i.e. 0.109%) and the "cost" of the initial assumption regarding class tempo probabilities (i.e. instances —outliers— that effectively have a tempo outside of their class's tempo range, i.e. 54 out of 698 instances).

This opens two important avenues for future work: improving the accuracy of the expert classifiers (for instance in refining the selection of the most relevant descriptors for each classifier) and study the validity of the limited-tempo-ranges assumption on a database containing more instances of a larger number of classes.

5.2 Content-based transformations

Transformations of audio signals have a long tradition (Zölzer, 2002). A recent trend in this area of research is the editing and transformation of musical audio signals triggered by explicit musically-meaningful representational elements, in contrast to low-level signal descriptors. These recent techniques have been coined content-based audio transformations (Amatriain et al., 2002), or adaptive digital audio effects (Verfaille et al., in press).

In this section, we describe a system for transformations of audio signals based on a description of their rhythmic structure.[19] The Swing Transformer consists in a content description module and a transformation module implying a high-quality time-scaling

---

[19]Part of the material in this section has been previously published in a conference article written with Lars Fabig and Jordi Bonada (Gouyon et al., 2003).

algorithm (Bonada, 2000). The former achieves an offline pre-analysis which achieves onset detection, determination of tempo and beat indexes at the quarter-note and eighth-note levels, as well as estimation of the swing ratio, if there is any. The transformation module consists in time-scaling of the audio in real-time. The time-scaling is controlled by a "User Swing Ratio." While playing back the audio file (in a loop), the user can continuously adjust the swing ratio in real-time: one can either increase or decrease the swing.

5.2.1 Swing estimation algorithm

Focusing on the swing of a musical excerpt requires the determination of two distinct metrical levels, a fast and a slow one. As swing is applied on eighth-notes, it is necessary to recognize which elements in the musical flow are eighth-notes. But this is not sufficient, one must also describe the excerpt at a higher (slower) metrical level. That is, determine the eighth-note "phase:" in a group of two eighth-notes, determine which is the first one. Indeed, it is not at all the same to perform a long-short pattern as a short-long pattern. The existing swing ratio (if there is any) must also be estimated.

Onset and transient detection Onsets are detected as on page 169. For later use in the time-scaling algorithm, transients are also extracted from the audio by an algorithm described in (Bonada, 2000). Here, the distinction "onset" vs "transient" is as follows: as detailed in (Gouyon et al., 2002), the rhythmic analysis input must consist in reliable note onsets. On the other hand, as explained in (Bonada, 2000), the time-scaling algorithm apply different processings on stable and transient regions, there, the detection of non-stationarities does not have to be restricted to note onsets. In one case (rhythmic analysis) the detection of non-stationarities should be rather oriented towards "no false-alarms", in the other case (time-scaling) it should rather be oriented towards "no missed."

IOI histogram computation IOIs are computed, taking into account the time differences between any two onsets. An IOI histogram is generated as on page 169.

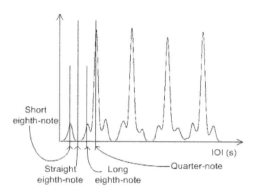

Figure 5.10: Example of an IOI histogram of an audio signal with a 2.7:1 swing ratio

As detailed below, the standard deviation of the Gaussian smoothing window is an important parameter. Then, peak positions and heights are detected in the histogram with an N-point running window method. One can verify on Figure 5.10 on this page and on Figure 5.11 on the next page the intuitive idea that peaks corresponding to shortened and lengthened eighth-notes are closer to the position of the straight eighth-note for small swing ratio (Figure 5.11) than for bigger ones (Figure 5.10). The estimation of the swing ratio relies on that observation.

Tatum period estimation   The tatum computation implements the assumption that music shows approximate integer timing ratios between metrical levels. IOI histograms should show peaks at approximately harmonic positions. Therefore, as on 170, the tatum is computed as the gap of the IOI histogram peak harmonic series. The use of the TWM procedure purposely filters out the deviations from exact integer ratios between pulses that occur e.g. in the case of music that swings. In this stage, the Gaussian standard variation is set to a medium value (e.g. 35 ms) not to be led astray by small IOIs.

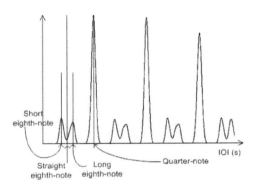

Figure 5.11: Example of an IOI histogram of an audio signal with a 1.5:1 swing ratio

Eighth-note and quarter-note period estimation   The quarter-note period is computed posterior to the tatum as the maximum peak in the IOI histogram among four candidates: the tatum, twice the tatum, three times the tatum and four times the tatum (with additional boundary restrictions: the quarter-note minimum tempo is set to 50 BPM, and the maximum to 270 BPM).

The eighth-note period is then computed as half the quarter-note.

Let us provide a justification of this procedure. The swing concerns the smallest metrical level present in the signal. As can be seen on Figure 5.10 on the preceding page and on Figure 5.11 on the current page, the amount of swing has a direct influence on the tatum estimation. Let us consider the following cases:

- If the audio has a swing ratio of 1:1 (i.e. "no swing"), the computation of the tatum yields the actual smallest level. That is, the tatum is the eighth-note.

- If the swing ratio is 2:1 ("ternary feel"), the IOI corresponding to the straight eighth-note is not present in the signal (nor in the histogram), there are solely shortened eighth-notes (whose durations are $\frac{1}{3}$ of that of a quarter-note) and lengthened eighth-notes (whose durations are $\frac{2}{3}$ of that of a quarter-note). There, the tatum computation yields $\frac{1}{3}$ of the quarter-note length.

- If the swing ratio is higher than 2:1 (above ternary feel), the IOI corresponding to the straight eighth-note is not present in the signal, there are solely shortened eighth-notes with durations smaller than $\frac{1}{3}$ of that of a straight quarter-note, and lengthened eighth-notes with durations superior to $\frac{2}{3}$ of that of a straight quarter-note. There, as it is restricted to integer ratios, the computation of the tatum yields either $\frac{1}{3}$ or $\frac{1}{4}$ of the quarter-note length.

Swing ratio estimation Two different implementations of the swing ratio estimation are still under tests. They are both based on the computation of a second IOI histogram, with a Gaussian standard deviation smaller than in the previous step (e.g. 10 ms), in order to account for more peaks and also a better time precision in the peak positions.

First implementation It is based on the computation of deviations between all peaks in the IOI histogram and integer multiples of the eight-note length. An important observation is that the deviation distribution is *bimodal*: one mode is around 0 (it corresponds to deviations with respect to quarter-note positions) and the second mode does correspond to relevant deviations for swing estimation (see Figure 5.12, in this example, the straight eighth-note length is 161ms, the deviation central tendency of the second mode is 68 ms; this results in a 2.45:1 swing ratio). The central tendency of the second mode deviations is computed (either as the mean, the median or the mode). Finally, the swing ratio is computed as:

$$Swing\ ratio = \frac{Eighth\ note\ period + central\ tendency}{Eighth\ note\ period - central\ tendency}$$

Second implementation The basic concept in this implementation is the seeking of the best match between IOI histogram peaks and swing templates. As can be seen in Figure 5.13, templates are built as harmonic comb grids with a gap equal to the quarter-note length and offsets varying between the eighth-note length for the first template and e.g. 3/4 of the quarter-note length for the last one (this is directly

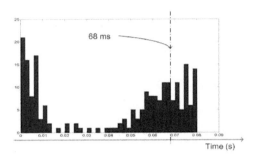

Figure 5.12: Distribution of the deviations between IOI histogram peaks and integer multiples of the eighth-note length.

related to the maximum boundary the user can set for swing ratio seeking). For each template, the TWM error function is computed between the grid elements and the IOI histogram peaks (see on page 170), then the best template is chosen as that which yields the smallest error (i.e. which best matches the peaks). This procedure resembles somehow the method proposed by Laroche (2001), the difference being that we do not achieve the matching over the onsets but the IOI histogram peaks.

Eighth-note and quarter-note position estimation Given the quarter-note period and the audio signal onsets, the quarter-note positions are sought (and quarter-note period is slightly adjusted) so as to match to the best the onset positions similarly as on page 170. Logically, each quarter-note position is also an eighth-note position. The remaining eighth-note positions are simply determined as positions in between each pair of quarter-notes: half-way between two quarter-notes, adjusted with respect to the detected swing ratio.

5.2.2 Swing transformations

Modifying the swing means moving onsets corresponding to eighth-notes from their original positions to different ones. In Figure 5.14, an example is shown for an audio file that has no swing (signal in the upper half of Figure 5.14). Quarter-notes are

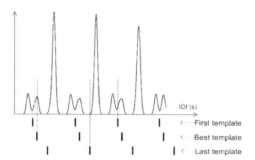

Figure 5.13: Illustration of the template-matching approach to swing estimation

depicted by a simple number ('1', '2', '3' on the figure top). The eighth-notes are indexed with i ($i = 1$ means "in a subdivision of a quarter-note in two eighth-notes, this is the first eighth-note", and $i = 2$ means "in a subdivision of a quarter-note in two eighth-notes, this is the second eighth-note") and their corresponding sample positions n_i. The detected Swing Ratio (SR) in the example is 1:1 (i.e. "no swing"). When the user chooses a different swing ratio (for example 2.6:1), the regions between indexes $n_{i=1}$ and $n_{i=2}$ are expanded with the time-scale factor TS_{EXP} while the regions between $n_{i=2}$ and $n_{i=1}$ are compressed with TS_{COMP}. The scaling factors for expansion and compression are calculated as follows ($TS > 1$ means signal expansion, and $TS < 1$, signal compression):

$$TS_{EXP} = \frac{SR_{User} + SR_{Detected} \cdot SR_{User}}{SR_{Detected} + SR_{Detected} \cdot SR_{User}}$$

$$TS_{COMP} = 1 + SR_{Detected}\,(1 - TS_{EXP})$$

When the original audio signal already has swing, the processing is slightly different because the regions of expansion and compression are not equal-sized anymore. The real onset positions of eighth-notes indexed by $i = 2$ deviate from the straight eighth-note grid. Regions to be expanded and to be compressed are adapted accordingly.

More details on the time-stretch algorithm can be found in (Bonada, 2000).

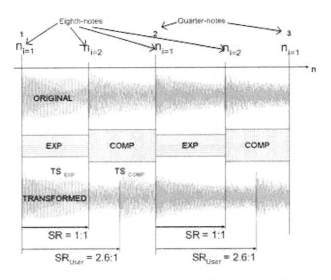

Figure 5.14: Adding swing to an audio file by time-scaling

5.2.3   Evaluation

The system provides very good sound quality for monophonic signals or polyphonic stereo mixes. It could be also extended to handle multi-channel signals.

Improvements are needed on the time-scale algorithm to reduce phasing and flanging for time-scale factors above 1.3 (note that this is a very high scaling factor). If we want to apply very drastic swing ratios, scaling factors higher than 1.3 are very often exceeded and sound quality may decrease. Although at large time-scale factors an analysis frame is repeated many times until the next frame is chosen, the resulting signal may sound metallic. This can be improved by interpolation of the magnitude spectrum between subsequent analysis frames. Another area for improvement is the handling of transients lying precisely on eighth-note positions. Probably, when the scaling factor is switched from expansion to compression (e.g. from 1.3 to 0.7) in a small group of frames belonging to a transient, this may cause a doubling of the transient (this is problematic especially for drum sounds).

Finally, it is our belief that the analysis of the deviation distribution should be further pursued. Indeed, the number of modes, the mode variances and higher moments (skewness and kurtosis) are probably representative of important information regarding diverse systematic timing deviations (Bilmes, 1993).

5.3   Conclusions

In this chapter, new algorithms have been presented for the automatic estimation from musical audio signals of diverse rhythmic descriptors: the tatum, time signature and swing. They are all grounded on the computation of rhythm periodicity functions. We also introduced other rhythmic descriptors of lower levels of abstraction, calculated from simple parameterizations of rhythm periodicity functions, such as coefficients inspired from the MFCCs. We illustrated the use of these descriptors in two music content processing scenarios: genre classification and rhythmic expressiveness transformations. We showed that tempo and tatum are relevant descriptors for genre classification. But, one the one hand, usual errors of automatic tempo induction algorithms (confusion of metrical levels) lower dramatically classification accuracy. And on the other hand, the assumption that tempo is genre-specific may not hold when considering very large number of musical pieces. This issue is left for future work. Several parameterizations of rhythm periodicity functions have been introduced in this chapter. Many, especially MFCC-like coefficients, yield good classification accuracy.

Experiments on the determination of time signature highlighted the relevance of the inter-beat segment temporal centroid in the determination of downbeats. Assuming that note occurrences have a direct correlation with increases in the waveform amplitude (and thus with the value of the temporal centroid), one might hypothesize that evidences for downbeats would be given by patterns of note timings. That is, the main difference between, on the one hand, a downbeat and the next beat and, on the other hand, an upbeat and the next beat would reside in the regularity of note timing patterns: along the musical sequence, patterns of note onset times would show greater similarity between a downbeat and its consecutive beat than between an upbeat and its consecutive beat. This hypothesis is an extension of the widespread hypothesis that the frequency of note occurrences would be greater on strong metrical time points; it would not be really that there are more notes between a downbeat and the next beat than between an upbeat and the next beat, but rather that these notes would show more regular patterns.

Finally, we demonstrated the concept of content-based transformation of audio signals by implementing a fully-automatic swing transformation algorithm that conserves sound quality.

Chapter 6

Conclusion

This book addressed several issues that appear when computers are asked to make diverse rhythmic responses to music as for instance "perceiving" tempo and beats. In this last chapter, we briefly summarize the contributions we believe this book makes to the research literature in computational rhythm description. Importantly, we believe that the main contributions are rather theoretical than practical and that the many questions raised in this book are more important to the improvement of current state-of-the-art that the algorithms proposed. Then, we embrace a slightly broader perspective and highlight things we did not try (but might have, if it wasn't for time constraints). We also propose lines of research that we believe are of very first interest but which we did not incorporate in our research either because they implied too distant methodologies or simply because we did not have a clear idea about how to do it.

6.1 Summary of contributions

Rhythm description functional framework The literature in automatic rhythm description is very furnished. However, we believe that no attempt had yet been made to provide a big picture of the functional blocks that existing systems have in common. In Figure 2.4, we depict an unifying functional framework for automatic rhythm

description which we believe permits to explain all existing systems as different instances of the same general model. In this framework, the computation of rhythm
periodicity function plays a central role.

Comprehensive review of rhythm description systems   Another important
contribution of this book is a comprehensive review and a qualitative comparison
of the rhythm description systems proposed in the literature with respect to the
functional units of the proposed framework (Section 2.2).

Evaluation of tempo induction algorithms   An important body of the literature
is dedicated to automatic tempo induction. As we have seen, current approaches tend
to deal directly with audio signals rather that symbolic signals or artificial sequences.
However, to our knowledge no attempt had yet been made to compare such algorithms
on a consequent data set.

As a first step towards more systematic evaluations and comparisons, we organized
a quantitative evaluation of some state-of-the-art algorithms for tempo induction
(Section 2.3) in the form of the first public benchmark on this topic (ran during
the International Conference on Music Information Retrieval held at the University
Pompeu Fabra in Barcelona in October 2004). In order to stimulate further research,
the contest results, annotations, evaluation software and part of the data are available
at http://ismir2004.ismir.net/ISMIR_Contest.html.

Current research directions   Via the qualitative comparisons of rhythm description systems and, additionally, the quantitative comparisons of a number of tempo
induction systems we provide a better understanding of recent achievements in automatic rhythm description, especially in tempo induction, and a clearer viewpoint on
current research directions and open issues.

As argued in more details in Section 2.4, one of the main observations is the
superiority of frame-based features over onsets for robust tempo induction.

With respect to open issues, we argue that improvements of tempo induction
systems depend on further research on the low-level features that summarize the

rhythmic aspects of musical data that further rhythmic processing deal with. More research is also needed in the design of rhythm periodicity functions and of methods for the combination and parsing of multiple rhythmic information sources. We believe that a crucial issue lies in the joint estimation of several metrical levels. Finally, much more effort should be dedicated to systematic evaluations of algorithms, for instance in the form of public competitions.

Low-level feature selection One of the open issues raised in this book concerns the selection of the low-level features that best summarize the rhythmic content of music. In Chapter 3, we propose an original methodology to address this problem, based on the assumption that low-level audio features that are adequate to the computational identification of beats are also appropriate for computing useful rhythmic periodicity functions, and thus for the rhythm description problem in general.

We conducted a series of experiments that led us to the following conclusions. Among the 274 features considered, on overall, the best individual feature is the variation of the energy. This confirms findings already found in the literature. However, we also showed that the notion of best individual feature depends on the musical genre considered. We showed that the association of several low-level features into feature sets (as for instance the variation of the energy in different frequency bands) results in a better description of the rhythmic content of music. Among the many possible combinations of features into sets, we motivated the use of a set of 59 features: the magnitude-normalized first-order differentials of the energy in ERB bands; the magnitude-normalized first-order differentials of the spectral peak harmonic centroid, spectral peak harmonic deviation, spectral peak mean, flatness, geometric mean, mean, kurtosis, skewness and low-frequency energy relation; the first-order differentials of the MFCCs and the phase deviation.

Echoing some open issues in the literature, we demonstrated the superiority of the ERB frequency subband decomposition over others (proposed in the literature) as the basis for the computation of effective energy feature sets. We also demonstrated that, for almost all feature subsets, the degree of change of feature magnitudes are better features than mere magnitudes. Normalizing a first-order magnitude differential by the magnitude seems to be the most accurate implementation for that. Another conclusion is that the energy in frequency regions below 500 Hz and above 5 kHz (approximatively) are more relevant than in mid-band frequencies.

Finally, we demonstrated the relevance of low-level features that current literature on rhythmic description does not consider, namely the Mel-Frequency Cepstrum Coefficients and a set of spectral features listed in Section 3.4.4.

Tempo induction algorithm   In Chapter 4, we addressed the specific problem of tempo induction from audio signals. We build algorithms based on the computation of the low-level feature sets advocated in a previous chapter. Following low-level feature computation, we considered several strategies to the computation of periodicity functions and the combination and parsing of multiple sources of information. Some algorithms show performances that are comparable to the state-of-the-art. We demonstrated that a special effort should be dedicated to the choice of algorithm input features. We also concluded on the need to consider constraints posed by the metrical hierarchy in the design of tempo induction algorithms. Most importantly, we illustrated the fact that the periodicity functions commonly used in tempo induction cannot fully take advantage of the explanatory power of large feature sets. More research is needed on this topic.

Also related to tempo induction, an important (and original) conclusion reached in Section 2.3 is that the implementation of robust tempo induction algorithms calls for the computation of low-level frame features rather than that of onset lists as the first processing block.

Tatum estimation algorithm   Having demonstrated in detail the use of rhythm periodicity functions in the task of tempo induction, we illustrated other applications for these functions, especially in the context of music content processing and Music Information Retrieval.

First, rhythm periodicity functions are useful for computing other pulses that the tactus. We implemented a working and reasonably reliable algorithm for tatum estimation based on a particular rhythm periodicity function (the inter-onset interval histogram) and the use of a pitch detection technique (the two-way mismatch) and demonstrated the use of the tatum in genre classification experiments. It should be noted that our tatum estimation algorithm was one of the very first published in the literature.

Time signature determination algorithm We also implemented a working and reliable algorithm for the automatic determination of the time signature of audio signals (as "duple" or "triple"). In this algorithm, the computation of a periodicity function also plays a crucial role. Here also, the publication of this algorithm represents one of the very first attempts to compute time signature directly from audio signals.

Swing estimation algorithm Also directly based on the continuous representation of a signal's periodicities, we proposed an algorithm for the estimation of swing in musical audio signals. Again, the algorithm is one the very first on its topic.

Rhythmic features for genre classification In addition to the tempo and the tatum, we introduced new rhythmic features for musical genre classification based on the parameterization of rhythmic periodicity functions. We demonstrated the relevance of these features and released the data sets used in our experiments in order to stimulate further research.[1]

In our opinion, our most original and valuable contribution to genre classification is the idea to compute usual spectral descriptors (as e.g. the flatness) and Mel-Frequency Cepstrum Coefficients on a rhythm periodicity function instead than on the Fourier transform of an audio signal. This had not been published before and had shown very good results.

Content-based transformation of audio We illustrated the concept of content-based transformation via a software application for rhythmic expressiveness transformations of musical audio signals. A fully automatic system has been developed, the

---

[1]http://www.iua.upf.es/mtg/ismir2004/contest/rhythmContest/

Swinger, which requires neither manual editing nor software or hardware sampler and provides very good sound quality, even for relatively extreme transformations.

6.2 Future work

6.2.1 Low-level rhythmic features

In Chapter 3, we considered many low-level features as front-end for the computation of useful periodicity functions. However, even if considerably larger than what is usually considered in current rhythm description literature, the list of features was limited. On the other hand, the literature on low-level audio features is very furnished. Further experiments could be conducted with for instance features extracted from models of the auditory system (as the low-level features proposed in the IPEM Matlab Toolbox),[2] diverse implementations of the loudness or MPEG7 features. Paragraph 3.4.3.1 concluded on the fact that the definition of the frequency subband decomposition for the computation of energy features is a sensitive factor. Some subbands perform better than others. This indicates that further work may also be dedicated to seeking the best subband definition. Features of a higher level of abstraction may also be considered in future work. For instance, features indicating the global pitch content of frames (Gómez and Herrera, 2004), or the presence of percussion instruments (Sandvold et al., 2004).

We have seen that robust tempo induction calls for the computation of low-level frame features rather than that of onset lists. However, whether this superiority of frame-based features over onset lists has any perceptual validity, as proposed by Scheirer (1998), remains to be investigated. Further, future work may be dedicated to challenging the perceptual validity of feature set rankings obtained in Chapter 3.

One might object that the methodology proposed in Chapter 3 may in fact characterize onsets as opposed to beats. Further work could be dedicated to repeat experiments of Chapter 3 with data labelled with onsets and non-onsets (instead of beats and non-beats) and compare the results with those detailed here.

---

[2]http://www.ipem.ugent.be/Toolbox/

6.2.2 Tempo induction

6.2.2.1 Periodicity functions

Experiments of Chapter 4 indicate that the autocorrelation function and comb filter-banks are better periodicity functions than the Fourier transform. The latter seems to perform slightly better than the former, however, no clear conclusion could be reached. Further, the performance of a given periodicity function also depends heavily on which strategies are used in the other blocks of a tempo induction algorithm. Therefore, more research should be dedicated to systematic evaluations of periodicity functions. Also, recall that experiments of Chapter 5 have been conducted with rhythm periodicity functions computed from audio features that were different from those advocated in Chapter 3. This is left for future work.

6.2.2.2 Combining and parsing multiple information sources

Chapter 4 detailed several methods for the combination of diverse sources of informations, together with their evaluations. We argued that an important algorithmic choice lies in information integration before or after periodicity function computation. That is, one either combines feature lists or periodicity functions. Future research could explore *cross-correlating* feature lists. This would probably emphasize common periodicities to all feature lists. Keeping the most prominent peaks would yield wining periodicities and (implicitly) winning features, offering a combined strategy for the evaluation of features, the computation of a periodicity function and the integration of multiple information sources.

We have seen that periodicity functions commonly used in tempo induction cannot fully take advantage of the explanatory power of *large* feature sets. A possible solution to this problem would be to include in the tempo induction algorithm a procedure that would further reduce the number of features depending on their individual behavior on *specific* pieces of music under analysis. This reduction of the number of features should be seen as complementary to the global feature selection strategy reported in Chapter 3. We proposed in Section 2.2.2.4 several methods for that. For instance, it is possible to measure the periodic, or non-periodic behavior of features and use this as a criterion to select among features. Preliminary experiments with the selection criteria exposed on page 46 have been conducted (they are not reported in this book), and did not give satisfying results. Another way to deal with this problem could be to consider replacing both the computation of periodicity functions and their subsequent combination by the use of pattern recognition techniques that would probably permit a more natural way to deal with both issues jointly and would probably be more effective in dealing with large numbers of features (see on page 214).

In Chapter 4, we considered several strategies for parsing multiple sources of information while others (detailed on page 49) have been left for future work. For instance, we have seen that the best parsing strategy among those used in Chapter 4 accounts for constraints posed by the metrical hierarchy ("musical parsing"). Other strategies should also be tested: e.g. seek periodicities in periodicity functions themselves,[3] compute feature lists at the scale of low metrical levels hence "going up" in the metrical hierarchy or consider rule-based or probabilistic frameworks to represent these constraints. Another strategy for periodicity function parsing is to keep several candidates (prominent peaks in the periodicity function) refine them through beat tracking and keep the one that propagates better, as proposed e.g. by Dixon (2001a). Experiments reported in Section 2.3 could not demonstrate the superiority of this method (implemented in DixonT) on the simpler highest-peak-picking method (implemented in DixonI). Here also, as in determining the relevance of multiplying periodicity functions with a tempo preference distribution (see page 24), more research and evaluations are needed.

6.2.2.3   Redundant approach to tempo induction

Section 2.3 shows several algorithms performing the same task and exhibiting specific performances on specific parts of the data. This raises an important question: Can we improve tempo estimation accuracy by combining the outputs of several algorithms?

The answer seems to be "yes,"[4] although it should be noted that simply computing

---

[3]Note that in former research, we concluded that this method leads to the induction of fast metrical levels, as the tatum, rather than the tactus.

[4]The original idea of "redundant approach" to tempo induction was originally formulated by Anssi Klapuri who did most of the analysis reported in this paragraph.

e.g. the median of the tempo estimates of different algorithms does *not* yield an improvement. This is because the "too slow" and "too fast" tempo estimates cannot be guaranteed to balance each other out.

A thorough analysis of algorithm skills and error trends would dictate a set of rules for combining algorithms. Lacking this information, one can think of a voting mechanism for combining the tempo estimates of different algorithms. Imagine different ordered lists of algorithms. Algorithms in the list being considered in turn, for each piece, a given algorithm gets one vote from all the algorithms that agree with its tempo estimate. An algorithm X is defined to agree with an algorithm Y if the ratio of their estimates is 1, 0.5, or 2, with 4% precision. The tempo estimate of the algorithm which gets the largest number of votes among the algorithms is selected as the output. If several algorithms receive the same number of votes, the order of the algorithms in the list defines which estimate is selected as the output.

An exhaustive search over all possible combinations of five algorithms (from among the 11) was conducted to find a combination which performs best using this voting mechanism. Applying the accuracy measure 1, the algorithms [Klapuri, Uhle, Klapuri, DixonI, DixonACF] achieved a performance of 68% and, applying accuracy measure 2, the algorithms [Klapuri, Scheirer, DixonT, DixonI, DixonACF] achieved a performance of 86%. This does not represent a significant improvement compared to the performance of Klapuri alone (67% and 84% according to the accuracy measures 1 and 2, respectively). However, the situation becomes clearer when Klapuri is excluded. In this case, the algorithms [Uhle, Scheirer, DixonI, DixonACF, DixonT] together achieve a rate of 57% with accuracy 1 and the algorithms [Scheirer, Uhle, DixonT, DixonACF, DixonACF] achieve a rate of 84% with accuracy 2. Compared to the best individual performances among the remaining algorithms (Uhle achieves 51% with accuracy 1 and DixonACF achieves 81% with accuracy 2), the voting mechanism makes a statistically significant improvement to the individual results.

The experiment described above is an example of a "redundant" approach to music content analysis: instead of designing one very complex algorithm we combine a number of different and more simple mechanisms. This idea stems from Bregman (1998), who pointed out that human perception appears to be redundant at many levels: there are several different processing principles serving the same purpose, and when one of them fails, another succeeds.

The combination of algorithms is an interesting avenue for future work and raises the following interesting questions: Which commonalities and differences should we implement in the concurring algorithms? How simple should we keep these algorithms? Is the voting scheme proposed above the best way to combine algorithms?

An interesting way to tackle this problem could be to embrace a machine learning perspective and focus on ensemble learning methods. In supervised learning, ensemble learning algorithms take decisions regarding the membership of a given instance to a class among several possible classes by considering the "votes" of several classifiers, previously trained on labeled data. Designing several classifiers for the same task can be done in several manners, for instance it is possible to provide different subsets of the training data to a base algorithm (as is done e.g. in *Bagging* and *Boosting*), it is also possible to provide the same training data but described with different attributes (Dietterich, 2002). To use these methods for tempo induction, one would need to define training and test data sets and, possibly, discretize tempo in a number of classes (see on page 215). Then, forcing diversity in the design of different algorithms could be done by specializing each of them on a restricted set of signal features (hence meeting the research direction proposed earlier on page 203). Another option could be to specialize algorithms to be combined on different categories of pieces.

6.2.2.4 Towards better benchmarks for algorithm evaluation

Section 2.3 details the first public large-scale cross-validation of audio tempo induction algorithms. Nevertheless, much effort is still needed to design better benchmarks for tempo induction algorithms. Here is a (non-exhaustive) list of issues that, in our opinion, should be addressed.

Beat tracking Tempo induction and beat tracking are part of the same perceptual process (Desain and Honing, 1999), future evaluation efforts should therefore consider them jointly.

Data More data is needed for future contests. Importantly, a larger amount of data with triple and other meters is required. However, not all music is suitable. As discussed and exemplified by Dixon (2001b), pieces can show diverse levels of difficulty. It may be difficult to induce the tempo or track the beats of specific musical pieces, even from constant-tempo performances, if they have a complex rhythmic structure (e.g. many events not on beats, or many beats occurring between musical events), while other pieces may be fundamentally less challenging. Additionally, in the case of performed music, keeping an almost steady tempo or adding expressive tempo variations is up to the performer. For instance, in Section 2.3, results are better on Ballroom data; this was predictable as this music was produced to ease learning of ballroom dances, hence has relatively clear beats and stable tempo.

Therefore, measuring the level of "rhythmic difficulty" of the pieces in the test set might provide an additional control for thorough evaluations. Goto and Muraoka (1997) and Dixon (2001b) propose such metrics.

Note on data annotation Gathering large sets of annotated data can be a time-consuming process. Further it is also prone to annotation errors. Therefore it is important to consider easing the task oh human annotators by for instance providing them adequate annotation tools. Few such tools exist for tempo and beat annotations.[5]

Goto and Muraoka (1997) refer to a "beat-position editor." This is a manual beat annotation tool that provides waveform visualization and, for accurate annotations, audio feedback in the form of short bursts of noise added on beats.

To our knowledge, the only publicly available softwares for semi-automatic beat annotation are BeatRoot (Dixon, 2001a) and the software reported in (Gouyon et al., 2004b). To lower the annotation effort, BeatRoot provides an automatic beat tracking algorithm in order to start (or continue) a bootstrapped interactive process of annotation between the user and the algorithm. Interactivity resides in that the user's corrections to the algorithm output (the beats) are fed back as inputs to the very

---

[5]The material in this paragraph was previously published in a stand-alone paper (Gouyon et al., 2004b). Coauthors of the paper are thanked for their collaboration.

algorithm.

In (Gouyon et al., 2004b), we report on a system built upon BeatRoot as well as other open source audio processing tools, namely WaveSurfer (Sjölander and Beskow, 2000) and CLAM.[6] The system is also publicly available.[7] Improvements upon Beat-Root are the following. The graphical interface shows important differences with BeatRoot. For instance, it is our belief that WaveSurfer's built-in visualization functionalities (e.g. running cursor and scrolling panes synchronized with audio playback), and its intuitive keyboard shortcuts and general look-and-feel are an enhancement of BeatRoot's interface. Also, an important point is that graphical configurations can be defined by the user. Other relevant differentiating features are the capture of keyboard messages while listening to the audio (hence permitting to annotate beats in real-time by tapping on a key while listening); the possibility to instantiate diverse transcription panes in order to annotate several metrical levels and database connection facilities (annotations can be stored locally and, when correct, they can easily be uploaded to a distant repository, for instance, a structured musical metadata database such as the MTG database (Cano et al., 2004)). However, it must be noted that BeatRoot also permits to annotate beats of MIDI data, which the system reported in (Gouyon et al., 2004b) does not permit.

Better annotations and evaluation measures   It is difficult to evaluate the accuracy of an algorithm for determining the correct tempo because of the inherent ambiguity of metrical levels: two listeners might not agree on a metrical level as the "correct" tempo.

Accuracy 2 (defined on page 70) was designed to account for this inherent ambiguity. However, its drawback (in our use of it) is that it does not take the meter into account. Considering half and double ground-truth tempo as correct makes sense solely for duple meter pieces. Similarly, considering three times and one third of ground-truth tempo as correct makes sense solely for pieces with a triple or compound meter.

---

[6]http://www.iua.upf.es/mtg/clam
[7]at http://www.iua.upf.es/~fgouyon/BeatTrackingPlugin/

In future contests, more accurate evaluations might be obtained by considering the "degree of ambiguity" of excerpt tempi. This could be done by recruiting several annotators (at least 10 or 15) for each piece and considering several metrical levels as valid options *only* in cases where annotators disagree on the tempo. This procedure could also tell us *which* levels are valid for each piece. A comparable procedure has been chosen by the organizers of the ISMIR 2005 tempo induction contest.[8] This is however a very time-consuming procedure.

A faster way to proceed to more precise evaluations would be to manually annotate beats of each piece at *2 different metrical levels* instead of one. Pieces would be annotated by a single person. For each piece, the accuracy measure of an algorithm would be the best match over the annotated metrical levels.

One might object that, for a given piece, two algorithms might not be evaluated with respect to the same metrical level. Nevertheless, both levels have been considered valid by the annotator. And we can assume that, in tempo-ambiguous cases, any two listeners would perceive at least one level in common, solely the rankings of metrical levels would differ. Consider the following example: a piece of music whose levels all share duple relationships, to which listener A taps the tempo at 50 BPM. Being asked to define another level, he chooses 100 BPM (it is highly unlikely that he would choose 25 BPM which is too slow to be a perceptually valid tempo). Say that listener B naturally taps the tempo at 200 BPM, being asked to define another level, he will most likely choose 100 BPM (not 400 BPM). Even in this extreme case, there exists some agreement. Thus, this procedure would be a way to measure *how close a specific algorithm gets to human agreement regarding tempo perception.* Such annotations could be done with the help of annotation tools as proposed e.g. by Dixon (2001a) and Gouyon et al. (2004b).

Robustness tests Other robustness tests are needed, for instance, robustness to increasing levels of noise (decreasing SNR) and robustness to cropping (the effect of the length of the excerpt).

---

[8]http://www.music-ir.org/mirexwiki/index.php/Audio_Tempo_Extraction

More modular evaluations   It is difficult to compare systems that, even if they implement similar concepts, do not share any piece of code. The performance of each system depends on the overall implementation and it is often hard to say anything more than "system A performed better than system B (on this data set)." That is, we are unable to say anything conclusive about the system submodules (for instance, whether frame differentials are better than absolute values), without being able to switch the submodules within a single system. On the other hand, it would be difficult to implement different systems in a common software framework so that they share simple processing blocks. Indeed, forcing the use of a specific implementation framework would probably have negative repercussions in terms of the number of entries in a competition. In the evaluations detailed in Section 2.3, different system variants from the same participant (Alonso, Dixon or Tzanetakis) give the most reliable information about the effect on the performance of different solutions for a given submodule of the system. A solution could therefore be to motivate participants of future competitions to submit several systems, with small, but conceptually relevant, variations in some submodules.

6.2.3   Rhythm as a tool for research in human audition

Music is present in all human societies and the perception and production of rhythm is ubiquitous in our lives (Desain and Windsor, 2000). An increasing number of researchers believe that the development of musical abilities during evolution helped the emergence of human cognitive abilities (Zatorre, 2005). Further, recent research argue in favor of the special role of musical *rhythm* in human evolution (Bispham and Cross, 2005). This leads to the conclusion that studying musical rhythm is another paradigm to the study of humans: music and rhythm provide excellent tools for investigating perceptual and cognitive processes involved in human audition.

Many theories and models of rhythm have been developed in diverse research communities (see on page 12), however, in addition to current open issues (summarized above in this chapter) in the design of working computational models, many fundamental questions remain. Consider the following ones for instance. Are there such things as "universal features" in rhythm perception? Which structures of the human brain are involved in the perception, recognition and memorization of rhythm patterns? In which balance do they imply high-level cognitive processes and sensory capabilities of lower level? Are these structures also involved in speech understanding, or motor control? In which way is rhythm perception linked with harmony, timbre and pitch perception? In which way does exposure to musical stimuli results in the emergence of internal representations and what is the influence of such representations in the audition of new stimuli? How do these structures evolve through life?

From the physics of musical sounds to the formation of abstract representations in the brain via the physiology of the auditory system, audition involves physical, physiological and cognitive aspects. Hence, it is our belief that a multidisciplinary approach is needed to reach a complete understanding of rhythm (Desain et al., 1998). Among the many possible methodologies to the study of rhythmic aspects of music (see on page 12), the work reported in this book focuses only on a computational methodology. Rhythm perception was addressed via a modeling approach implying signal processing and machine learning methods. In addition to improvements of the methods detailed in this book, future work should also be oriented towards complementary approaches, both analytical (cognitive psychology and neurophysiology) and computational (computational neuroscience and robotics). The analytical approach can confirm or refute implementation principles of artificial auditory systems and can also provide insights for the design of new models (as artificial neural networks (Husain et al., 2004)).

Final goals of this multidisciplinary research are, on the one hand, a better understanding of the cognitive processes involved in human audition and on the other hand the design of artificial systems with cognitive abilities that would participate in a more intuitive integration of technology in most our daily activities. The remainder of this section details diverse research paths towards these goals.

6.2.3.1 Networks of rhythmic units

We will follow a computational approach to the study of cerebral functional modules involved in the perception of rhythm, putting a particular emphasis on the study of large networks of simple rhythmic units. Different types of simple units will be studied, as e.g. those inspired by physical oscillators (Large and Kolen, 1994; McAuley, 1995) and spiking neurons (Eck, 2002b). We will first investigate the properties of single units when fed with simple rhythmic inputs (Eck, 2002a). We will then study the propagation of patterns of large networks of oscillators decoupled from musical inputs, explore the properties of different network topologies (Cano et al., 2005a), varying e.g. the connectivity between units —i.e. who is connected to whom— and explore synchronization phenomena between them (Strogatz, 2003; Neda et al., 2000, 2003). Such networks will finally be connected to external musical acoustic stimuli (rather than simplified artificial sequences or MIDI inputs, as is usual in the literature on oscillator models for rhythm perception) and a special focus will be put on the computation and selection of relevant acoustic features as inputs to these networks. A link will be made with results of the work proposed on page 202.

6.2.3.2 Localization of rhythmic areas in the brain

Recent progress has been made on the determination of neural areas related to time estimation (Coull et al., 2004). We will follow a line of research focused on the estimation of *musical* time. Under the hypothesis that there exist areas in our brains responsible for music processing and, at a more detailed level, for e.g. pitch and rhythm processing (Peretz and Coltheart, 2003; Peretz and Zatorre, 2005), we will begin collaborations with neuroscientists to tackle the problem of localization of such areas via brain imagery methods (Pfeuty et al., 2003; Janata and Grafton, 2003; Desain, 2004; Levitin and Menon, 2005).

A first experimental protocol will be based on the detection of modules specialized in tempo *changes* rather than in general rhythm processing. Control musical sequences will be generated with diverse tempo curves (using for instance state-of-the-art time-scaling algorithms (Bonada, 2000)) and transmitted to subjects whose brain activity will be recorded via magnetic resonance imaging for instance. Correlations will then be sought between the control tempo curves and recordings of areas activated while listening to the stimuli.

A better knowledge of the brain areas involved in tempo perception would be of interest in the current debate regarding whether it involves complex high-level cognitive processes and memory retrieval or whether it rather involves physiological processes of lower levels of cognition (Todd, 1994; Scheirer, 1998). This, in turn, would provide insights to the determination of the required complexity of computational models.

6.2.3.3 Active rhythm perception

The perception of rhythm is very much linked to rhythm *production*: it is common, even for non-musicians, to produce physical rhythmic responses such as foot tapping, swinging, dancing or even simply adapting one's gait, in phase with external acoustic stimuli.

With this argument in mind, we will follow a reactive approach to the modeling of rhythm perception and embrace Brooks' "behavior-based" approach to Artificial Intelligence (1991a; 1991b).

Brooks is a researcher in robotics who argued (with others, e.g. (Mataric, 1997; Steels, 1999; Pfeifer and Scheier, 1999)) in favor of physical embodiment in the modeling of cognitive processes. He criticizes the "traditional" view of Artificial Intelligence that places high-level reasoning and symbolic representations as central, and critical, modules where cognition would take place and that would receive information *from* sensor systems and send information *towards* actuation systems. In his approach, sensors and actuators are directly coupled (i.e. actually wired through a series of *continuous* transformational stages), there is no central component dedicated to cognition, cognition would only be in the eye of the observer who would "attribute cognitive abilities to a system that works well in the world but has no explicit place where cognition is done." For Brooks, cognition *emerges* from the interaction of a "creature" with its environment. A creature should therefore be situated in the world (i.e. cope in real-time with the dynamic and "dirty" environment), have a purpose (*do something*) and finally be itself part of the world (its actions have immediate effects on its own sensations.

Applications of this rationale in the rhythm domain can be found in (Eck et al., 2000), (Bryson, 1992) and (Franklin, 2001). Other mentions to this rationale can be found in (Scheirer, 2000, p.74), (Iyer, 2002) and (Gouyon and Meudic, 2003).

Also following recent sensory-motor theories of perception (Todd et al., 2002; O'Regan and Noë, 2001), we will implement interactive rhythmic systems based on a strong coupling between rhythm perception and production functional modules. This will provide a good domain for the study of interactions between the human brain, the body and the environment.

6.2.3.4   Learning by specialization

As early as 1987, Dannenberg and Mont-Reynaud (1987) suggested that procedural approaches would be better suited to music understanding than declarative approaches: "[...] it seems that music understanding relies more heavily on pattern recognition capabilities [...] than it does on logical reasoning and problem-solving techniques. It is possible that an emphasis on the latter is precisely the privilege of musical experts, who are able to name, and to reason about, entities which naive listeners also perceive, for the most part, but are unable to put into words."

We have seen that recent efficient computational models of rhythm description implement problem solving techniques and probabilistic approaches. State variables (e.g. beats, quantized durations) are considered hidden variables to be estimated from observations. Efficient models have been implemented (e.g. (Rosenthal, 1992; Cemgil and Kappen, 2003; Raphael, 2002)) to search wide hypothesis spaces. In this powerful framework, knowledge is not declarative in the sense that it would explicitly be constrained by fixed, deterministic sets of rules (as those reviewed in (Desain and Honing, 1999)), however, it does lie in probabilistic models learned from measured data during a supervised training step.

In connection to the reactive approach to rhythm perception advocated above, we will explore the idea that human categorization of musical stimuli in categories such as tempo classes or rhythm patterns entails both the retrieval of pre-constructed mental representations as well as a strong influence of present context.[9] In terms of machine learning, this would mean for instance that general models could first be

---

[9]and therefore follow some of our previous work in this direction (Gouyon, 2000; Sandvold et al., 2004).

constructed, via supervised learning strategies using large collections of musical items, or generic rules could be formulated. These representations would subsequently be progressively adapted (through unsupervised learning) to *each* analyzed item.

Consider for instance a general model of tempo induction accounting for diverse low-level features and procedures to seek temporal recurrences of such features. While analyzing a particular piece of music, some features may reveal more relevant than others and should therefore be given more importance. Different features would very likely be relevant for other pieces. An example of rhythm pattern determination using clustering techniques posterior to a metrical analysis can also be found in (Dixon et al., 2004).

6.2.3.5 Tempo ambiguity

We have seen that it is common to assign to pieces of music a scalar value, its tempo in BPM, supposed to be representative of its speed. However, we have also seen in Section 2.1.2 that some psychological research asserts that the perception of tempo is an ambiguous phenomenon, ambiguity referring here to the rate of agreement among listeners. There does not seem to exist *one objective tempo* to attribute to a piece of music.

Following studies by Moelants and McKinney (2004) and McKinney and Moelants (2004), we will collect tapping data from a sufficiently large number of subjects and determine in which balance tempo values are consensual or ambiguous. The goal will then be to investigate which low-level signal features (i.e. *objective* features) correlate with the distribution of listeners' responses. In sum, we will explore to which extent tempo perception is relative to the signal itself. Hopefully, this experiment will result in an algorithm to measure tempo ambiguity directly on the signal.

6.2.3.6 Perceptual tempo categories

As already mentioned, the ambiguity of the very concept of tempo makes the evaluation of tempo induction algorithm difficult, even when using carefully annotated data sets. The design of evaluation metrics that would take this into account is still the object of current research (see on page 206 and the ISMIR 2005 tempo induction contest proposal for evaluation metrics).[10]

It may be interesting to approach the notion of tempo of a piece of music not as a scalar in BPM, but rather as a category among a discrete set of possible ones as for instance "very slow," "slow," "fast" and "very fast," or perhaps the most common Italian tempo markings as found on many scores ("Largo," "Adagio," "Lento," "Andante," "Moderato," "Allegretto," "Allegro" and "Presto"). The availability of annotated data could favor certain categories over others.

Machine learning methods, fed by signal attributes as periodicity functions, selected peak thereof or for instance mean IOI, complexity of the metrical hierarchy and event density as proposed by Drake et al. (1999), could be used in the design of models of tempo categories.

---

[10]http://www.music-ir.org/mirexwiki/index.php/Audio_Tempo_Extraction

Bibliography

Aigrain, P. (1999). New applications of content processing of music. *Journal of New Music Research*, 28(4):271–280.

Alghoniemy, M. and Tewfik, A. (1999). Rhythm and periodicity detection in polyphonic music. In *Proc. IEEE Workshop on Multimedia Signal Processing*, pages 185–190. IEEE Signal Processing Society.

Allen, P. and Dannenberg, R. (1990). Tracking musical beats in real time. In *Proc. International Computer Music Conference*, pages 140–143.

Alonso, M., David, B., and Richard, G. (2004). Tempo and beat estimation of musical signals. In *Proc. International Conference on Music Information Retrieval*, pages 158–163, Barcelona. Audiovisual Institute, Universitat Pompeu Fabra.

Amatriain, X., Bonada, J., Loscos, A., and Serra, X. (2002). Spectral processing. In Zölzer, U., editor, *DAFX Digital Audio Effects*. J. Wiley and Sons.

Aucouturier, J.-J. and Pachet, F. (2003). Representing musical genre: A state of the art. *Journal of New Music Research*, 32(1):83–94.

Baggi, L. (1991). Neurswing: An intelligent workbench for the investigation of swing in jazz. *Computer*, 24(7):60–64.

Bello, J. (2003). *Towards the Automated Analysis of Simple Polyphonic Music: A Knowledge-based Approach*. PhD Thesis, Dept of Electronic Engineering, Queen Mary University of London, London.

217

Bello, J., Duxbury, C., Davies, M., and Sandler, M. (2004). On the use of phase and energy for musical onset detection in the complex domain. *IEEE Signal Processing Letters*, 11(6):553–556.

Bello, J. and Sandler, M. (2003). Phase-based note onset detection for music signals. In *Proc. IEEE International Conference on Acoustics, Speech, and Signal Processing*. IEEE Signal Processing Society.

Bilmes, J. (1993). *Timing is of the Essence: Perceptual and Computational Techniques for Representing, Learning, and Reproducing Expressive Timing in Percussive Rhythm*. Master Thesis, MIT, Cambridge.

Bispham, J. and Cross, I. (2005). Homo sapiens sapiens: The rhythmic species. In *10th Rhythm Perception and Production Workshop*.

Blum, A. and Langley, P. (1997). Selection of relevant features and examples in machine learning. *Artificial Intelligence*, 97(1-2):245–271.

Blum, T., Keislar, D., Wheaton, A., and Wold, E. (1999). Method and article of manufacture for content-based analysis, storage, retrieval, and segmentation of audio information. USA Patent 5,918,223.

Bonada, J. (2000). Automatic technique in frequency domain for near- lossless timescale modification of audio. In *Proc. of the International Computer Music Conference*, pages 396–399.

Bregman, A. (1998). Psychological data and computational ASA. In Rosenthal, D. and Okuno, H., editors, *Computational Auditory Scene Analysis*. Lawrence Erlbaum Associates, New Jersey.

Brooks, R. (1991a). Intelligence without representation. *Artificial Intelligence Journal*, 47:139–159.

Brooks, R. (1991b). New approaches to robotics. *Science*, 253:1227–1232.

Brown, J. (1993). Determination of the meter of musical scores by autocorrelation. *Journal of the Acoustical Society of America*, 94(4):1953–1957.

Brown, J. and Puckette, M. (1989). Calculation of a narrowed autocorrelation function. *Journal of the Acoustical Society of America*, 85(4):1595–1601.

Bryson, J. (1992). *The Subsumption Strategy Development of a Music Modelling System*. Master Thesis, University of Edinburgh, Faculty of Science, Department of Artificial Intelligence, Edinburgh.

Cambouropoulos, E., Dixon, S., Goebl, W., and Widmer, G. (2001). Computational models of tempo: Comparison of human and computer beat-tracking. In *Proc. International Symposium on Systematic and Comparative Musicology and International Conference on Cognitive Musicology*, pages 18–26.

Cano, P., Celma, O., Koppenberger, M., and Martin-Buldú, J. (2005a). The topology of music artists' graphs. In *Proc. XII Congreso de Física Estadística*.

Cano, P., Koppenberger, M., Ferradans, S., Martinez, A., Gouyon, F., Sandvold, V., Tarasov, V., and Wack, N. (2004). MTG-DB: A test environment for music audio processing. In *Proc. International Conference on Web Delivering of Music*.

Cano, P., Koppenberger, M., Wack, N., G. Mahedero, J., Masip, J., Celma, O., Garcia, D., Gómez, E., Gouyon, F., Guaus, E., Herrera, P., Massaguer, J., Ong, B., Ramírez, M., Streich, S., and Serra, X. (2005b). An industrial-strength content-based music recommendation system. In *Proc. 28th ACM International Conference on Research and Development in Information Retrieval*.

Cemgil, A., Desain, P., and Kappen, B. (2000). Rhythm quantization for transcription. *Computer Music Journal*, 24(2):60–76.

Cemgil, A. and Kappen, B. (2001). Tempo tracking and rhythm quantization by sequential monte carlo. In A., C. and B., K., editors, *Proc. Advances in Neural Information Processing Systems*.

Cemgil, A. and Kappen, B. (2003). Monte carlo methods for tempo tracking and rhythm quantization. *Journal of Artificial Intelligence Research*, 18:45–81.

Cemgil, A., Kappen, B., Desain, P., and Honing, H. (2001). On tempo tracking: Tempogram representation and Kalman filtering. *Journal of New Music Research*, 28(4):259–273.

Chowning, J., Rush, L., Mont-Reynaud, B., Chafe, C., Schloss, A., and Smith, J. (1984). Intelligent system for the analysis of digitized acoustic signals. Report STAN-M-15, CCRMA, Stanford University, Palo Alto.

Chung, J. (1989). *An agency for the perception of musical beats or If I had a foot...* Master Thesis, MIT, Cambridge.

Clarke, E. (1987). Levels of structure in the organization of musical time. *Contemporary music review*, 2(1):211–238.

Clarke, E. (1999). Rhythm and timing in music. In Deutsch, D., editor, *The Psychology of Music*, Series in Cognition and Perception. Academic Press, 2nd edition.

Clynes, M. and Walker, J. (1982). Neurobiologic functions of rhythm, time, and pulse in music. In Clynes, M., editor, *Music, mind, and brain: The neuropsychology of music*, pages 171–216. Plenum.

Clynes, M. and Walker, J. (1986). Music as time's measure. *Music Perception*, 4(1):85–119.

Cooper, G. and Meyer, L. (1960). *The rhythmic structure of music*. University of Chicago Press, Chicago.

Coull, J., Vidal, F., Nazarian, B., and Macar, F. (2004). Functional anatomy of the attentional modulation of time estimation. *Science*, 303:1506–1508.

Dahl, S. (2005). *On the beat - Human movement and timing in the production and perception of music*. PhD Thesis, KTH Royal Institute of Technology, Stockholm.

Dannenberg, R. and Mont-Reynaud, B. (1987). Following an improvisation in real-time. In *Proc. International Computer Music Conference*, pages 241–258.

Dash, M. and Liu, H. (1997). Feature selection for classification. *Intelligent Data Analysis*, 1:131–156.

Desain, P. (1992). A (de)composable theory of rhythm perception. *Music Perception*, 9(4):439–454.

Desain, P. (2004). What rhythm do i have in mind? detection of imagined temporal patterns from single trial erp. In *Proc. 8th International Conference on Music Perception and Cognition*, page 209.

Desain, P. and de Vos, S. (1990). Autocorrelation and the study of musical expression. In *Proc. International Computer Music Conference*, pages 357–360.

Desain, P. and Honing, H. (1989). The quantization of musical time: A connectionist approach. *Computer Music Journal*, 13(3):55–66.

Desain, P. and Honing, H. (1991). Tempo curves considered harmful. A critical review of the representation of timing in computer music. In *Proc. International Computer Music Conference*, pages 143–149.

Desain, P. and Honing, H. (1999). Computational models of beat induction: The rule-based approach. *Journal of New Music Research*, 28(1):29–42.

Desain, P., Honing, H., van Thienen, H., and Windsor, L. (1998). Computational modeling of music cognition: Problem or solution? *Music Perception*, 16(1):151–166.

Desain, P. and Windsor, L., editors (2000). *Rhythm Perception and Production*. Swets and Zeitlinger.

Dietterich, T. (1998). Approximate statistical tests for comparing supervised classification learning algorithms. *Neural Computation*, 10(7):1895–1924.

Dietterich, T. (2002). Ensemble learning. In Arbib, M., editor, *The Handbook of Brain Theory and Neural Networks*. MIT Press, Cambridge MA, 2nd edition.

Dixon, S. (1999). A beat tracking system for audio signals. In *Proc. Conference on Mathematical and Computational Methods in Music (Diderot)*, pages 101–110.

Dixon, S. (2001a). Automatic extraction of tempo and beat from expressive performances. *Journal of New Music Research*, 30(1):39–58.

Dixon, S. (2001b). An empirical comparison of tempo trackers. In *Proc. 8th Brazilian Symposium on Computer Music*, pages 832–840.

Dixon, S. (2005). Perceptual smoothness of tempo in expressively performed music. *Music Perception*, in press.

Dixon, S. and Cambouropoulos, E. (2000). Beat tracking with musical knowledge. In *Proc. European Conference on Artificial Intelligence*, pages 626–630. IOS Press, Amsterdam.

Dixon, S., Gouyon, F., and Widmer, G. (2004). Towards characterisation of music via rhythmic patterns. In *Proc. International Conference on Music Information Retrieval*, pages 509–516, Barcelona. Audiovisual Institute, Universitat Pompeu Fabra.

Dixon, S., Pampalk, E., and Widmer, G. (2003). Classification of dance music by periodicity patterns. In *Proc. International Conference on Music Information Retrieval*, pages 159–165.

Downie, J. S. (2003). Music information retrieval. *Annual Review of Information Science and Technology*, 37:295–343.

Drake, C. (1993). Reproduction of musical rhythms by children, adult musicians and adult non-musicians. *Perception and Psychophysics*, 53(1):25–33.

Drake, C. and Bertrand, D. (2001). The quest for universals in temporal processing of music. *Annals of the New York Academy of Science*, 930:17–27.

Drake, C., Gros, L., and Penel, A. (1999). How fast is that music? The relation between physical and perceived tempo. In Yi, S., editor, *Music, Mind and Science*. Seoul National University Press.

Drake, C., Jones, M., and Baruch, C. (2000a). The development of rhythmic attending in auditory sequences: attunement, referent period, focal attending. *Cognition*, 77(3):251–288.

Drake, C., Penel, A., and Bigand, E. (2000b). Why musicians tap slower than nonmusicians. In Desain, P. and Windsor, L., editors, *Rhythm Perception and Production*, pages 245–248. Swets and Zeitlinger.

Duda, R., Hart, P., and Stork, D. (2001). *Pattern classification*. J. Wiley and Sons, New York, 2nd edition.

Duxbury, C., Sandler, M., and Davies, M. (2002). A hybrid approach to musical note onset detection. In *Proc. Digital Audio Effects Conference*, pages 33–38, Hamburg. University of the German Federal Armed Forces.

Eck, D. (2001). A network of relaxation oscillators that finds downbeats in rhythms. In Dorffner, G., editor, *Proc. International Conference on Artificial Neural Networks*, pages 1239–1247, Berlin. Springer.

Eck, D. (2002a). Finding downbeats with a relaxation oscillator. *Psychol. Research*, 66(1):18–25.

Eck, D. (2002b). Real-time musical beat induction with spiking neural networks. Technical Report IDSIA-22-02, IDSIA.

Eck, D., Gasser, M., and Port, R. (2000). Dynamics and embodiment in beat induction. In Desain, P. and Windsor, L., editors, *Rhythm Perception and Production*, pages 157–170. Swets and Zeitlinger.

Foote, J. (2000). Automatic audio segmentation using a measure of audio novelty. In *Proc. IEEE Conference on Multimedia and Expo*, pages 452–455.

Foote, J., Cooper, M., and Nam, U. (2002). Audio retrieval by rhythmic similarity. In *Proc. International Conference on Music Information Retrieval*, pages 265–266.

Foote, J. and Uchihashi, S. (2001). The Beat Spectrum: A new approach to rhythm analysis. In *Proc. International Conference on Multimedia and Expo*, pages 881–884.

Fraisse, P. (1982). Rhythm and tempo. In Deutsch, D., editor, *The Psychology of Music*, Series in Cognition and Perception, book chapter 6, pages 149–180. Academic Press.

Franklin, J. (2001). Multi-phase learning for jazz improvisation and interaction. In *Proc. Eighth Biennial Symposium on Arts and Technology (Perception and Interaction in the Electronic Arts)*.

Friberg, A. and Sundberg, J. (1995). Time discrimination in a monotonic, isochronous sequence. *Journal of the Acoustical Society of America*, 98(5):2524–2531.

Friberg, A. and Sundström, J. (1999). Jazz drummers' swing ratio in relation to tempo. In *Proc. Acoustical Society of America ASA/EAA/DAGA Meeting Lay Language Papers*.

Friberg, A. and Sundström, J. (2002). Swing ratios and ensemble timing in jazz performances: Evidence for a common rhythmic pattern. *Music Perception*, 19(3):333–349.

Gabrielsson, A. (1973a). Similarity ratings and dimension analyses of auditory rhythm patterns. Part I. *Scandanavian Journal of Psychology*, 14:138–160.

Gabrielsson, A. (1973b). Similarity ratings and dimension analyses of auditory rhythm patterns. Part II. *Scandanavian Journal of Psychology*, 14:161–176.

Gasser, M., Eck, D., and Port, R. (1999). Meter as mechanism: A neural network that learns metrical patterns. *Connection Science*, 11(2):187–216.

Gillick, L. and Cox, S. (1989). Some statistical issues in the comparison of speech recognition algorithms. In *Proc. IEEE Conference on Acoustics, Speech and Signal Processing*, pages 532–535.

Gómez, E. (2005). Tonal description of polyphonic audio for music content processing. *INFORMS Journal on Computing*, in press.

Gómez, E. and Herrera, P. (2004). Automatic extraction of tonal metadata from polyphonic audio recordings. In *Proc. 25th International AES Conference*, pages 74–80.

Gómez, E., Klapuri, A., and Meudic, B. (2003). Melody description and extraction in the context of music content processing. *Journal of New Music Research*, 32(1):23–41.

Goto, M. (2001). An audio-based real-time beat tracking system for music with or without drums. *Journal of New Music Research*, 30(2):159–171.

Goto, M. and Muraoka, Y. (1995). A real-time beat tracking system for audio signals. In *Proc. International Computer Music Conference*.

Goto, M. and Muraoka, Y. (1997). Issues in evaluating beat-tracking systems. In *Proc. International Joint Conferences on Artificial Intelligence, Workshop on Computational Auditory Scene Analysis*, pages 9–16.

Goto, M. and Muraoka, Y. (1999). Real-time beat tracking for drumless audio signals: Chord change detection for musical decisions. *Speech Communication*, 27(3-4):311–335.

Gouyon, F. (2000). *Extraction automatique de descripteurs rythmiques dans des extraits de musiques populaires polyphoniques.* Diplôme d'Etudes Approfondies Thesis & Internal report, IRCAM, Centre Georges Pompidou, Paris & Sony CSL, Paris.

Gouyon, F. (2003). *Towards Automatic Rhythm Description of Musical Audio Signals. Representations, Computational Models and Applications.* Pre-doctoral Thesis, Universitat Pompeu Fabra, Barcelona.

Gouyon, F., Amatriain, X., Bonada, J., Cano, P., Gómez, E., Herrera, P., and Loscos, A. (in press). Content processing of musical audio signals. In Leman, L. and Cirotteau, D., editors, *Sound to sense, sense to sound: A state-of-the-art*.

Gouyon, F. and Dixon, S. (2004). Dance music classification: A tempo-based approach. In *Proc. International Conference on Music Information Retrieval*, pages 501–504, Barcelona. Audiovisual Institute, Universitat Pompeu Fabra.

Gouyon, F. and Dixon, S. (2005). A review of automatic rhythm description systems. *Computer Music Journal*, 29(1):34–54.

Gouyon, F., Dixon, S., Pampalk, E., and Widmer, G. (2004a). Evaluating rhythmic descriptors for musical genre classification. In *Proc. 25th International AES Conference*, pages 196–204, London. Audio Engineering Society.

Gouyon, F., Fabig, L., and Bonada, J. (2003). Rhythmic expressiveness transformations of audio recordings: Swing modifications. In *Proc. International Conference on Digital Audio Effects*, pages 94–99, London.

Gouyon, F. and Herrera, P. (2003a). A beat induction method for musical audio signals. In Izquierdo, E., editor, *Proc. 4th European Workshop on Image Analysis for Multimedia Interactive Services*, pages 281–287, Singapore. World Scientific Publishing.

Gouyon, F. and Herrera, P. (2003b). Determination of the meter of musical audio signals: Seeking recurrences in beat segment descriptors. In *Proc. 114th AES Convention*. Audio Engineering Society.

Gouyon, F., Herrera, P., and Cano, P. (2002). Pulse-dependent analyses of percussive music. In *Proc. 22nd International AES Conference*, pages 396–401. Audio Engineering Society.

Gouyon, F., Klapuri, A., Dixon, S., Alonso, M., Tzanetakis, G., Uhle, C., and Cano, P. (2006). An experimental comparison of audio tempo induction algorithms. *IEEE Trans. Speech and Audio Processing*, in press.

Gouyon, F. and Meudic, B. (2003). Towards rhythmic content processing of musical signals - Fostering complementary approaches. *Journal of New Music Research*, 32(1):41–65.

Gouyon, F., Pachet, F., and Delerue, O. (2000). On the use of zero-crossing rate for an application of classification of percussive sounds. In *Proc. Digital Audio Effects conference*.

Gouyon, F., Wack, N., and Dixon, S. (2004b). An open-source tool for semi-automatic rhythmic annotation. In *Proc. International Conference on Digital Audio Effects*, pages 193–196.

Guyon, I. and Elisseeff, A. (2003). An introduction to variable and feature selection. *Journal of Machine Learning Research*, 3:1157–1182.

Hainsworth, S. (2004). *Techniques for the Automated Analysis of Musical Audio*. PhD Thesis, Cambridge University, Engineering Department, Cambridge.

Hainsworth, S. and Macleod, M. (2003). Beat tracking with particle filtering algorithms. In *Proc. IEEE Worshop on Applications of Signal Processing to Audio and Acoustics*, pages 91–94.

Hainsworth, S. and Macleod, M. (2004). Particle filtering applied to musical tempo tracking. *EURASIP Journal on Applied Signal Processing*, 15:2385–2395.

Hall, M. and Holmes, G. (2003). Benchmarking attribute selection techniques for discrete class data mining. *IEEE Transactions on Knowledge and Data Engineering*, 15(6):1437–1447.

Heittola, T. and Klapuri, A. (2002). Locating segments with drums in music signals. In *Proc. International Conference on Music Information Retrieval*, pages 271–272, Paris. IRCAM.

Herre, J., Cremer, M., Uhle, C., and Rohden, J. (2002). Proposal for a core experiment on AudioTempo. Report MPEG2001/8415.

Herrera, P., Dehamel, A., and Gouyon, F. (2003). Automatic labeling of unpitched percussion sounds. In *Proc. 114th AES Convention.* Audio Engineering Society.

Hofmann-Engl, L. (2002). Rhythmic similarity: A theoretical and empirical approach. In *Proc. International Conference on Music Perception and Cognition.*

Honing, H. (1993). Issues in the representation of time and structure in music. *Contemporary music review*, 9:221–239.

Honing, H. (2001). From time to time: The representation of timing and tempo. *Computer Music Journal*, 25(3):50–61.

Honing, H. (2005). Timing is tempo-specific. In *Proc. International Computer Music Conference*, pages 359–362.

Husain, F., Tagamets, M., Fromm, S., Braun, A., and Horwitz, B. (2004). Relating neuronal dynamics for auditory object processing to neuroimaging activity: A computational modeling and an fmri study. *Neuroimage*, 21(4):1701–1720.

Iyer, V. (1998). *Microstructures of Feel, Macrostructures of Sound: Embodied Cognition in West African and African-American Musics.* PhD Thesis, University of California, Berkeley.

Iyer, V. (2002). Embodied mind, situated cognition, and expressive microtiming in African-American music. *Music Perception*, 19(3):387–414.

Jain, A., Duin, R., and Mao, J. (2000). Statistical pattern recognition. *IEEE Transactions on Pattern Analysis and Machine Intelligence*, 22(1):4–37.

Janata, P. and Grafton, S. (2003). Swinging in the brain: Shared neural substrates for behaviors related to sequencing and music. *Nature Neuroscience*, 6:682–687.

Jones, M. and Boltz, M. (1989). Dynamic attending and responses to time. *Psychological Review*, 96(3):459–491.

Klapuri, A. (1999). Sound onset detection by applying psychoacoustic knowledge. In *Proc. IEEE International Conference on Acoustics, Speech and Signal Processing.*

Klapuri, A. (2004). Automatic music transcription as we know it today. *Journal of New Music Research*, 33(3):269–282.

Klapuri, A., Eronen, A., and Astola, J. (2005). Analysis of the meter of acoustic musical signals. *IEEE Trans. Speech and Audio Processing*, in press.

Kohavi, R. and John, G. (1997). Wrappers for feature selection. *Artificial Intelligence*, 97(1-2):273–324.

Lapidaki, E. (1996). *Consistency of tempo judgments as a measure of time experience in music listening*. PhD Thesis, Northwestern University, Evanston.

Lapidaki, E. (2000). Stability of tempo perception in music listening. *Music Education Research*, 2(1):25–44.

Large, E. and Kolen, E. (1994). Resonance and the perception of musical meter. *Connection Science*, 6:177–208.

Large, E. and Palmer, C. (2002). Perceiving temporal regularity in music. *Cognitive Science*, 26:1–37.

Laroche, J. (2001). Estimating tempo, swing and beat locations in audio recordings. In *Proc. IEEE Workshop on Applications of Signal Processing to Audio and Acoustics*, pages 135–138.

Laroche, J. (2003). Efficient tempo and beat tracking in audio recordings. *Journal of the Acoustical Society of America*, 51(4):226–233.

Lartillot, O. (2004). *Fondements d'un système d'analyse musicale computationnelle suivant une modélisation cognitiviste de l'écoute*. PhD Thesis, Université Paris VI, Paris.

Lerdahl, F. and Jackendoff, R. (1983). *A generative theory of tonal music*. MIT Press, Cambridge.

Levitin, D. and Cook, P. (1996). Absolute memory for musical tempo: Additional evidence that auditory memory is absolute. *Perception and Psychophysics*, 58(6):927–935.

Levitin, D. and Menon, V. (2005). The neural locus of temporal structure and expectancies in music: Evidence from functional neuroimaging at 3 Tesla. *Music Perception*, 22(3):563–575.

Liu, H. and Yu, L. (2005). Toward integrating feature selection algorithms for classification and clustering. *IEEE Transactions on Knowledge and Data Engineering*, 17(4):491–502.

London, J. (2005). Rhythm. In Macy, L., editor, *Grove Music Online*. Oxford University Press, http://www.grovemusic.com (Accessed 23 April 2005).

Longuet-Higgins, C. (1987). *Mental processes*. MIT Press, Cambridge.

Longuet-Higgins, C. and Lee, C. (1982). Perception of musical rhythms. *Perception*, 11:115–128.

Madison, G. (2000). On the nature of variability in isochronous serial interval production. In Desain, P. and Windsor, L., editors, *Rhythm Perception and Production*, pages 95–114. Swets and Zeitlinger.

Madison, G. and Merker, B. (2002). On the limits of anisochrony in pulse attribution. *Psychological Research*, 66(3):201–207.

Maher, J. and Beauchamp, J. (1993). Fundamental frequency estimation of musical signals using a two-way mismatch procedure. *Journal of the Acoust. Soc. of America*, 95(4):2254–2263.

Mataric, M. (1997). Studying the role of embodiment in cognition. *Cybernetics and Systems (special issue on Epistemological Aspects of Embodied AI)*, 28(6):457–470.

McAuley, J. (1995). *Perception of time as phase: Towards an adaptive-oscillator model of rhythmic pattern processing*. PhD Thesis, Indiana University, Bloomington.

McAuley, J. and Semple, P. (1999). The effect of tempo and musical experience on perceived beat. *Australian Journal of Psychology*, 51(3):176–187.

McKinney, M. and Moelants, D. (2004). Deviations from the resonance theory of tempo induction. In *Proc. Conference on Interdisciplinary Musicology*.

Meudic, B. (2002). Automatic meter extraction from MIDI files. In *Proc. Journées d'informatique musicale*.

Moelants, D. (2002). Preferred tempo reconsidered. In Stevensa, C., Burnham, D., McPherson, G., Schubert, E., and Renwick, J., editors, *Proc. of the International Conference on Music Perception and Cognition*, pages 580–583, Sydney.

Moelants, D. (2003). Dance music, movement and tempo preferences. In *Proc. 5th Triennal ESCOM Conference*.

Moelants, D. and McKinney, M. (2004). Tempo perception and musical content: What makes a piece fast, slow or temporally ambiguous. In *Proc. International Conference on Music Perception and Cognition*.

Mont-Reynaud, B. and Goldstein, M. (1985). On finding rhythmic patterns in musical lines. In *Proc.International Computer Music Conference*, pages 391–397.

Moore, B. (1995). *Hearing - Handbook of perception and cognition*. Academic Press, Inc., London, 2nd edition.

Neda, Z., Nikitin, A., and Vicsek, T. (2003). Synchronization of two-mode stochastic oscillators: a new model for rhythmic applause and much more. *Physica A*, 321(1-2):238–247.

Neda, Z., Ravasz, E., Brechet, Y., Vicsek., T., and Barabasi, A.-L. (2000). The sound of many hands clapping. *Nature*, 403:849–850.

Oppenheim, A. and Schafer, R. (2004). From frequency to quefrency: A history of the cepstrum. *IEEE Signal Processing Magazine*, 21(5):95–106.

O'Regan, J. and Noë, A. (2001). A sensorimotor account of vision and visual con-
sciousness. *Behavioral and Brain Sciences*, 24(5):939–1011.

Palmer, C. (1997). Music performance. *Annual review of psychology*, 48:115–138.

Pampalk, E., Dixon, S., and G., W. (2003). Exploring music collections by browsing
different views. In *Proc. International Conference on Music Information Retrieval*,
pages 201–208.

Pampalk, E., Rauber, A., and Merkl, D. (2002). Content-based organization and visualization of music archives. In *Proc. ACM International Conference on Multimedia*,
pages 570–579.

Parncutt, R. (1994). A perceptual model of pulse salience and metrical accent in
musical rhythms. *Music Perception*, 11(4):409–464.

Paulus, J. and Klapuri, A. (2002). Measuring the similarity of rhythmic patterns.
In *Proc. International Conference on Music Information Retrieval*, pages 150–156.
IRCAM - Centre Pompidou.

Peretz, I. and Coltheart, M. (2003). Modularity of music processing. *Nature Neuroscience*, 6(7):688–690.

Peretz, I. and Zatorre, R. (2005). Brain organization for music processing. *Ann. Rev.
Psychol.*, 56:89–114.

Pfeifer, R. and Scheier, C. (1999). *Understanding intelligence*. MIT Press, Cambridge.

Pfeuty, M., Ragot, R., and V., P. (2003). Processes involved in tempo perception: A
CNV analysis. *Psychophysiology*, 40:69–76.

Povel, D. and Essens, P. (1985). Perception of temporal patterns. *Music Perception*,
2(4):411–440.

Quinlan, R. (1993). *C4.5: Programs for Machine Learning*. Morgan Kaufmann, San
Mateo.

Raphael, C. (2002). A hybrid graphical model for rhythmic parsing. *Artificial Intelligence*, 137(1-2):217–238.

Repp, B. (1992). Probing the cognitive representation of musical time: Structural constraints on the perception of timing perturbations. *Cognition*, 44:241–281.

Repp, B. (1994). On determining the basic tempo of an expressive music performance. *Psychology of Music*, 22:157–167.

Rosenthal, D. (1992). *Machine rhythm: Computer emulation of human rhythm perception*. PhD Thesis, MIT, Cambridge.

Sandvold, V., Gouyon, F., and Herrera, P. (2004). Percussion classification in polyphonic audio recordings using localized sound models. In *Proc. International Conference on Music Information Retrieval*, pages 537–540, Barcelona. Audiovisual Institute, Universitat Pompeu Fabra.

Scheirer, E. (1997). Pulse tracking with a pitch tracker. In *Proc. IEEE Worshop on Applications of Signal Processing to Audio and Acoustics*.

Scheirer, E. (1998). Tempo and beat analysis of acoustic musical signals. *Journal of the Acoustical Society of America*, 103(1):588–601.

Scheirer, E. (2000). *Music-listening systems*. PhD Thesis, MIT, Cambridge.

Scheirer, E. and Slaney, M. (1997). Construction and evaluation of a robust multi-feature speech/music discriminator. In *Proc. IEEE-ICASSP*, pages 1331–1334.

Schloss, A. (1985). *On the automatic transcription of percussive music - From acoustic signal to high-level analysis*. PhD Thesis, CCRMA, Stanford University, Palo Alto.

Seppänen, J. (2001). *Computational models of musical meter recognition*. Master Thesis, Tampere University of Technology, Tampere.

Serra, X. (1989). *A system for sound analysis/transformation/synthesis based on a deterministic plus stochastic decomposition*. PhD Thesis, CCRMA Stanford University, Palo Alto.

Sethares, W., Morris, R., and Sethares, J. (2005). Beat tracking of musical performances using low-level audio features. *IEEE Trans. Speech and Audio Processing*, 13(2):275–285.

Sethares, W. and Staley, T. (2001). Meter and periodicity in musical performance. *Journal of New Music Research*, 30(2):149–158.

Sjölander, K. and Beskow, J. (2000). Wavesurfer - An open source speech tool. In *Proc. International Conference on Spoken Language Processing*.

Smith, L. (1996). Modelling rhythm perception by continuous time- frequency analysis. In *Proc. International Computer Music Conference*.

Smith, L. and Kovesi, P. (1996). A continuous time-frequency approach to representing rhythmic strata. In *Proc. International Conference on Music Perception and Cognition*.

Snyder, J. and Krumhansl, C. (2001). Tapping to ragtime: Cues to pulse finding. *Music Perception*, 18(4):455–489.

Steels, L. (1999). *The Talking Heads Experiment - Volume I. Words and Meanings*. Pre-edition for Laboratorium, Antwerpen.

Strogatz, S. (2003). *Sync: The emerging science of spontaneous order*. Hyperion, New York.

Tanguiane, A. (1993). *Artificial Perception and Music Recognition*. Springer, Berlin.

Tanguiane, A. (1994). A principle of correlativity of perception and its applications to music recognition. *Music Perception*, 11(4):465–506.

Temperley, D. (2004). An evaluation system for metrical models. *Computer Music Journal*, 28(3):28–44.

Temperley, D. and Sleator, D. (1999). Modeling meter and harmony: A preference-rule approach. *Computer Music Journal*, 23(1):10–27.

Thiemel, M. (2005). Accent. In Macy, L., editor, *Grove Music Online*. Oxford University Press, http://www.grovemusic.com (Accessed 23 April 2005).

Thornburg, H. (2001a). Bayesian segmentation and rhythm tracking. Part I: Constructing and identifying probabilistic models of rhythm. CCRMA report, Stanford University, http://ccrma.stanford.edu/~harv23/Thornburg_rt_pres1.pdf (Accessed 26 April 2005).

Thornburg, H. (2001b). Bayesian segmentation and rhythm tracking. Part II: Segmentation with priors. CCRMA report, Stanford University, http://ccrma.stanford.edu/~harv23/Thornburg_rt_pres2.pdf (Accessed 26 April 2005).

Thornburg, H. and Gouyon, F. (2000). A flexible analysis/synthesis method for transients. In *Proc. International Computer Music Conference*, pages 400–403.

Tillmann, B., Bharucha, J., and Bigand, E. (in press). Music perception from a connectionist perspective. In Zatorre, R. and Peretz, I., editors, *The Biological Foundations of Music*. Oxford University Press.

Todd, P. (1994). The auditory primal sketch: A multiscale model of rhythmic grouping. *Journal of New Music Research*, 23(1):25–70.

Todd, P., Lee, C., and Boyle, D. (2002). A sensory-motor theory of beat induction. *Psychological Research*, 66(1):26–39.

Tzanetakis, G. and Cook, P. (2002). Musical genre classification of audio signals. *IEEE Trans. Speech and Audio Processing*, 10(5):293–302.

Uhle, C., Rohden, J., Cremer, M., and Herre, J. (2004). Low complexity musical meter estimation from polyphonic music. In *Proc. AES 25th International Conference*, pages 63–68. Audio Engineering Society.

Vercoe, B. (1997). Computational auditory pathways to music understanding. In Del ège, I. and Sloboda, J., editors, *Perception and Cognition of Music*, pages 307–326. Psychology Press.

Verfaille, V., Zölzer, U., and Arfib, D. (in press). Adaptive digital audio effects (A-DAFx): A new class of sound transformations. *IEEE Trans. Speech and Audio Processing*.

Waadeland, C. (2001). It don't mean a thing if it ain't got that swing - Simulating expressive timing by modulated movements. *Journal of New Music Research*, 30(1):23–37.

Wang, Y. and Vilermo, M. (2001). A compressed domain beat detector using MP3 audio bitstreams. In *Proc. ACM Multimedia*.

Witten, I. and Frank, E. (2000). *Data Mining: Practical machine learning tools with Java implementations*. Morgan Kaufmann, San Francisco.

Yeston, M. (1976). *The stratification of musical rhythm*. Yale University Press, New Haven.

Yoshii, K., Goto, M., and Okuno, H. (2004). Automatic drum sound description for real-world music using template adaptation and matching methods. In *Proc. International Conference on Music Information Retrieval*, pages 184–191, Barcelona. Audiovisual Institute, Universitat Pompeu Fabra.

Zatorre, Z. (2005). Music, the food of neuroscience? *Nature*, 434:312–315.

Zils, A., Pachet, F., Delerue, O., and Gouyon, F. (2002). Automatic extraction of drum tracks from polyphonic music signals. In *Proc. International Conference on Web Delivery of Music*.

Zölzer, U., editor (2002). *DAFX Digital Audio Effects*. J. Wiley and Sons.

Appendix A

Detailed results of Chapter 4

In this appendix, we provide details of the performance of some algorithms taking oart in the experiments of Chapter 4.

	All		**Ballroom**		**Loops**		**Alonso**	
	acc1	acc2	acc1	acc2	acc1	acc2	acc1	acc2
set 1	45.7	75.7	54.6	84.1	43.5	71.6	42.3	81
set 2	42.4	67.1	39.8	66.6	44.4	65	37.8	77.1
set 3	54.7	**84.9**	63.5	88.1	53.9	82.9	45.8	88.5
set 4	63.8	83.4	61.6	83.2	67.3	**83.4**	51.9	**87.9**
set 5	62.6	82.4	62	81.4	66.4	82.4	48	84
set 6	46.4	83.7	56.6	90.5	44.6	80.8	39.1	85.9
set 7	44.2	80.9	55.1	90.1	42.9	78.4	33.7	78.3
set 8	48.5	82	58.4	85.7	48.7	80.6	33.3	82.4
set 9	57.2	75	48	64.2	62.9	79.4	46.6	72.2
set 10	49.7	80.1	57.2	83.2	50.1	79.4	37.6	78.7
set 11	56.7	83.6	69.2	88	54	81.2	50.3	87.1
set 12	53.5	82.3	57	86.5	54.8	80.8	42.9	82.6
set 13	46.3	84.7	57.9	**91.3**	44.2	82.1	38.6	86.5
set 14	48.4	84.5	60.6	89.4	47.3	82.8	35.4	84.9

Table A.1: Accuracies (in %) of some feature sets associated with algorithm Algorithm 12 for tempo induction. Bold fonts are used to highlight the best feature set given a data set (i.e. best line given a column)

	All		Ballroom		Loops		Alonso	
	acc1	acc2	acc1	acc2	acc1	acc2	acc1	acc2
set 1	29.5	62.3	30.7	66	28.1	59.6	33.3	68.1
set 2	25.2	51.9	17.8	42.5	26.3	52.1	30.9	64.2
set 3	27.4	63	28.2	67.9	25.7	58.9	33.5	73.2
set 4	23.3	61.4	23.1	67.3	21.2	57.1	32.3	71
set 5	24.3	62.4	25.2	**70.3**	22.4	58.4	30.7	67.7
set 6	28.6	63.7	28.5	68	27.6	61	33.1	68.9
set 7	29	63.7	29.2	68.5	28.2	60.5	32.7	70.5
set 8	26.1	62.9	25.1	68.5	25.1	59.7	31.5	68.1
set 9	22.6	54.4	16.3	47.8	23.8	55	26.2	60.1
set 10	26.5	60.1	25.8	65.6	26.1	57.3	28.8	64
set 11	28.1	62.8	27.5	69.8	26.8	57.9	34.7	**73.4**
set 12	25.7	59.8	27.1	60.5	24.2	58.4	29.9	65
set 13	28.9	**64.5**	28.8	69.8	27.4	**61.1**	35.6	71.2
set 14	28	64	28.2	69.6	26.1	60.1	35.2	71.4

Table A.2: Accuracies (in %) of some feature sets associated with algorithm Algorithm 4 for tempo induction. Bold fonts are used to highlight the best feature set given a data set (i.e. best line given a column)

	All		Ballroom		Loops		Alonso	
	acc1	acc2	acc1	acc2	acc1	acc2	acc1	acc2
set 1	29.5	62.3	30.7	66	28.1	59.6	33.3	68.1
set 2	25.2	51.9	17.8	42.5	26.3	52.1	30.9	64.2
set 3	29	60.9	30.1	66	27.6	57.1	33.3	69.7
set 4	24	60.8	25.4	67.8	22.5	56.7	28.2	67.9
set 5	25.7	60.5	27.9	66.8	24.1	58.2	29.6	61.1
set 6	29.8	63.2	30.7	66.8	29.3	59.9	30.7	72.2
set 7	29.8	63.3	29.7	67.2	29.2	60.2	32.5	70.7
set 8	23.4	56.2	19.3	59.2	24.2	53.8	26.2	61.5
set 9	21.4	46.3	13.5	36.5	23.9	48	22.5	53
set 10	26.2	57.2	23.3	61	26.8	55.2	27.4	60.3
set 11	29.4	62.1	27.6	**68.6**	28.7	57.2	35	**73**
set 12	25.1	57.9	25.8	59.5	23.5	55.8	30.9	64.2
set 13	30.8	**63.8**	30.5	68.3	29.9	**60.4**	34.8	71.8
set 14	27	59.1	23.6	65.2	26.9	55.4	32.1	65.8

Table A.3: Accuracies (in %) of some feature sets associated with algorithm Algorithm 1 for tempo induction. Bold fonts are used to highlight the best feature set given a data set (i.e. best line given a column)

	All		Ballroom		Loops		Alonso	
	acc1	acc2	acc1	acc2	acc1	acc2	acc1	acc2
set 1	16.4	49	38.8	86.1	7.4	31.6	22.3	68.3
set 2	15.2	46.5	35.1	79.7	6.8	30.1	21.9	67.3
set 3	17.8	50.5	43.8	90.5	7	32	25.6	70.5
set 4	15.5	35.6	34.4	76.5	6.3	17.2	27	53.8
set 5	17.3	41.2	42.4	83.5	6.3	21.8	27.4	61.8
set 6	17.2	50.4	41.7	90.7	7.4	31.6	22.7	**70.8**
set 7	17.7	50.2	41.3	90.5	8	31.5	24.7	70.5
set 8	18.4	**50.8**	44.4	**92.1**	7.8	**31.9**	25.6	**70.8**
set 9	12.7	44	30.4	74.5	6.6	30.9	13.1	55
set 10	16.9	47.4	41.4	89.3	6.9	29	23.5	64.4
set 11	18.4	48.9	44.7	90.3	7.2	29.8	27.6	69.3
set 12	15.7	47.1	36.2	81.7	7.3	30.9	21.5	65
set 13	17.5	42.8	42.5	85	5.9	22.2	29.9	68.1
set 14	17.4	42.6	42.4	84.8	5.8	22	30.3	68.1

Table A.4: Accuracies (in %) of some feature sets associated with algorithm Algorithm 6 for tempo induction. Bold fonts are used to highlight the best feature set given a data set (i.e. best line given a column)

	All		Ballroom		Loops		Alonso	
	acc1	acc2	acc1	acc2	acc1	acc2	acc1	acc2
set 1	16.4	49	38.8	86.1	7.4	31.6	22.3	68.3
set 2	15.2	46.5	35.1	79.7	6.8	30.1	21.9	67.3
set 3	17.1	50.2	41.1	88.8	7.2	32.3	24.1	69.5
set 4	17	42.7	36	80.7	7.8	24.5	28.4	64.6
set 5	17.8	46	40.7	83.4	7.8	27.9	26.8	65.2
set 6	15.8	49.4	38.8	87	6.5	32	22.1	67.9
set 7	16.8	49.5	40	88.1	6.9	31.5	24.7	69.5
set 8	17.6	46.1	44.6	90.3	6.7	26.8	24.1	63.2
set 9	12.8	40.3	29.9	72.6	6.4	26.5	15.1	51.5
set 10	16.8	46.1	43	87.2	6.9	27.4	21.9	65
set 11	17.9	**50.9**	43.3	90.3	6.9	**32.5**	27.4	**71.4**
set 12	15.7	46.3	36	81.2	7.5	29.8	21.1	64.8
set 13	17.1	49.8	42.3	91.4	6.8	31	23.9	68.5
set 14	18.1	47.6	46.6	**92.7**	6.8	27.6	24.7	66.5

Table A.5: Accuracies (in %) of some feature sets associated with algorithm Algorithm 3 for tempo induction. Bold fonts are used to highlight the best feature set given a data set (i.e. best line given a column)

	All		Ballroom		Loops		Alonso	
	acc1	acc2	acc1	acc2	acc1	acc2	acc1	acc2
set 1	29.5	62.3	30.7	66	28.1	59.6	33.3	68.1
set 2	25.2	51.9	17.8	42.5	26.3	52.1	30.9	64.2
set 3	26.4	60.9	27.2	66.9	24.6	56.1	32.3	**72.6**
set 4	22.1	58.8	23.5	66.5	19.9	53.9	29.4	68.1
set 5	22.9	59.1	23.9	68.6	21	54.5	29	64.8
set 6	27.6	62.1	28.4	**69.3**	25.6	57.5	34.8	71
set 7	29.3	**63.5**	29.4	68.8	28.3	**59.9**	33.3	71
set 8	26	62.8	24.5	68.3	25.2	59.5	31.5	68.3
set 9	8.2	21.2	0.7	2.7	12.4	31.7	1.2	4.1
set 10	25.4	58	22.1	60	26	56.5	27.8	61.8
set 11	26.8	60.2	25.1	66.5	26.2	55.6	31.5	70.8
set 12	25.1	59.4	26.4	60.5	23.6	57.6	29.6	65.4
set 13	26.6	60.7	25.5	64.2	25.3	57.4	33.5	69.5
set 14	26.5	61.8	23.3	62.7	25.7	59.2	34.4	71.2

Table A.6: Accuracies (in %) of some feature sets associated with algorithm Algorithm 7 for tempo induction. Bold fonts are used to highlight the best feature set given a data set (i.e. best line given a column)

	All		Ballroom		Loops		Alonso	
	acc1	acc2	acc1	acc2	acc1	acc2	acc1	acc2
set 1	16.4	49	38.8	86.1	7.4	31.6	22.3	68.3
set 2	15.2	46.5	35.1	79.6	6.8	30.1	21.9	67.3
set 3	17.5	50.4	41	89.4	7.5	32.3	25.8	70.1
set 4	16.8	40.2	33.8	78.9	8	21.5	29.6	63
set 5	17.1	43.4	37.2	83.4	7.8	24.8	26.8	63.8
set 6	16.8	49.1	40.1	88.5	7.2	31	23.7	68.5
set 7	17.1	49.1	40.5	88.8	7.3	30.8	24.3	68.7
set 8	17.9	47.7	44.3	90.8	7.2	28.6	24.9	65.6
set 9	12.7	43.7	30.2	73.5	6.4	30.5	14.1	56
set 10	16.7	45.8	41.3	89	6.7	26.7	23.3	63.6
set 11	18.5	**50.6**	42.5	90.4	7.9	**32.1**	28.4	71.2
set 12	15.4	46.8	35.2	80.2	7.1	30.8	21.3	65.8
set 13	18	**50.6**	42.1	91	7.9	31.7	25.6	**71.4**
set 14	18.8	49	44.8	**92**	7.8	29.3	27.6	70.1

Table A.7: Accuracies (in %) of some feature sets associated with algorithm Algorithm 9 for tempo induction. Bold fonts are used to highlight the best feature set given a data set (i.e. best line given a column)

Appendix B

Related publications by the author

In this annex, we provide a list of publications of relevance to this book in which its author has participated. Abstracts and electronic versions of most of these publications are available from http://www.iua.upf.es/mtg.

Journal articles

- **Authors**: Gouyon, F. Klapuri, A. Dixon, S. Alonso, M. Tzanetakis, G. Uhle, C. Cano, P.
  Title: An experimental comparison of audio tempo induction algorithms
  Journal: IEEE Transactions on Speech and Audio Processing, in press
  Year: 2006
  Related to Chapter 2

- **Authors**: Gouyon, F. Dixon, S.
  Title: A review of automatic rhythm description systems
  Journal: Computer Music Journal, 29(1)
  Year: 2005
  Related to Chapter 2

- **Authors**: Gouyon, F. Meudic, B.
  Title: Towards Rhythmic Content Processing of Musical Signals: Fostering

Complementary Approaches
Journal: Journal of New Music Research, 32(1)
Year: 2003
Related to Chapters 2 and 5

Book chapters

- **Authors**: Gouyon, F. Amatriain, X. Bonada, J. Cano, P. Gómez, E. Herrera, P. Loscos, A.
  Title: Content processing of musical audio signals
  Book title: Sound to sense, sense to sound: A state-of-the-art
  Editor: Leman, L. Cirotteau, D.
  Year: in press
  Related to Chapter 5

Theses and reports

- **Authors**: Gouyon, F.
  Title: Towards Automatic Rhythm Description of Musical Audio Signals — Representations, Computational Models and Applications
  Type: Pre-doctoral Thesis
  Institution: Universitat Pompeu Fabra, Barcelona
  Year: 2003
  Related to Chapters 2, 3 and 5

- **Authors**: Gouyon, F.
  Title: Extraction de descripteurs rythmiques dans des extraits de musiques populaires polyphoniques
  Type: DEA ATIAM Thesis & internal report
  Institution: IRCAM Centre Georges Pompidou, Paris & Sony CSL Labs, Paris
  Year: 2000

Related to Chapters 2, 3 and 5

- **Authors**: Gouyon, F.
  Title: Detection and Modeling of transient regions in musical signals
  Type: DEA SIC Thesis & internal report
  Institution: ENSEEIHT, Toulouse & CCRMA, Stanford University
  Year: 1999
  Related to Chapter 2

Presentations in conferences

- **Authors**: Gouyon, F. Widmer, G. Serra, X.
  Title: Acoustic Cues to Beat Induction: A Machine Learning Perspective
  Conference: 10th Rhythm Perception and Production Workshop
  Year: 2005
  Related to Chapter 3

- **Authors**: Cano, P. Koppenberger, M. Wack, N. Garcia, J. Masip, J. Celma, O. Garcia, D. Gómez, E. Gouyon, F. Guaus, E. Herrera, P. Massaguer, J. Ong, B. Ramírez, M. Streich, S. Serra, X.
  Title: An Industrial-Strength Content-based Music Recommendation System
  Conference: 28th ACM International Conference on Research and Development in Information Retrieval
  Year: 2005
  Related to Chapter 5

- **Authors**: Gouyon, F. Dixon, S. Pampalk, E. Widmer, G.
  Title: Evaluating rhythmic descriptors for musical genre classification
  Conference: 25th International AES Conference
  Year: 2004
  Related to Chapter 5

- **Authors**: Gouyon, F. Dixon, S.
  Title: Dance music classification: A tempo-based approach
  Conference: International Conference on Music Information Retrieval
  Year: 2004
  Related to Chapter 5

- **Authors**: Gouyon, F. Wack, N. Dixon, S.
  Title: An open source tool for semi-automatic rhythmic annotation
  Conference: International Conference on Digital Audio Effects
  Year: 2004
  Related to Chapters 2 and 5

- **Authors**: Dixon, S. Gouyon, F. Widmer, G.
  Title: Towards Characterisation of Music via Rhythmic Patterns
  Conference: International Conference on Music Information Retrieval
  Year: 2004
  Related to Chapter 5

- **Authors**: Cano, P. Koppenberger, M. Ferradans, S. Martinez, A. Gouyon, F. Sandvold, V. Tarasov, V. Wack, N.
  Title: MTG-DB: A Repository for Music Audio Processing
  Conference: International Conference on Web Delivering of Music
  Year: 2004
  Related to Chapter 5

- **Authors**: Sandvold, V. Gouyon, F. Herrera, P.
  Title: Drum sound classification in polyphonic audio recordings using localized sound models
  Conference: International Conference on Music Information Retrieval
  Year: 2004
  Related to Chapter 3

- **Authors**: Herrera, P. Sandvold, V. Gouyon, F.
  Title: Percussion-related Semantic Descriptors of Music Audio Files

Conference: 25th International AES Conference
Year: 2004
Related to Chapter 5

- **Authors**: Celma, O. Gómez, E. Janer, J. Gouyon, F. Herrera, P. García, D.
  Title: Tools for Content-Based Retrieval and Transformation of Audio Using MPEG7: The SPOffline and the MDTools
  Conference:
  Year: 25th International AES Conference
  Related to Chapter 5

- **Authors**: Cano, P. Fabig, L. Gouyon, F. Koppenberger, M. Loscos, A. Barbosa, A.
  Title: Semi-Automatic Ambiance Generation
  Conference: 7th International Conference on Digital Audio Effects
  Year: 2004
  Related to Chapter 5

- **Authors**: Gouyon, F. Herrera, P.
  Title: A beat induction method for musical audio signals
  Conference: 4th WIAMIS-Special session on Audio Segmentation and Digital Music
  Year: 2003
  Related to Chapter 4

- **Authors**: Gouyon, F. Herrera, P.
  Title: Determination of the Meter of musical audio signals: Seeking recurrences in beat segment descriptors
  Conference: Audio Engineering Society, 114th Convention
  Year: 2003
  Related to Chapter 5

- **Authors**: Gouyon, F. Fabig, L. Bonada, J.
  Title: Rhythmic expressiveness transformations of audio recordings: swing

modifications
Conference: 6th International Conference on Digital Audio Effects
Year: 2003
Related to Chapter 5

- **Authors**: Herrera, P. Dehamel, A. Gouyon, F.
  Title: Automatic Labeling of Un-pitched Percussion Sounds
  Conference: Audio Engineering Society, 114th Convention
  Year: 2003
  Related to Chapter 3

- **Authors**: Gómez, E. Gouyon, F. Herrera, P. Amatriain, X.
  Title: MPEG7 for Content-based Music Processing
  Conference: 4th WIAMIS-Special session on Audio Segmentation and Digital Music
  Year: 2003
  Related to Chapter 5

- **Authors**: Gómez, E. Gouyon, F. Herrera, P. Amatriain, X.
  Title: Using and enhancing the current MPEG7 standard for a music content processing tool
  Conference: Audio Engineering Society, 114th Convention
  Year: 2003
  Related to Chapter 5

- **Authors**: Gouyon, F. Herrera, P. Cano, P.
  Title: Pulse-dependent analyses of percussive music
  Conference: AES 22nd International Conference on Virtual, Synthetic and Entertainment Audio
  Year: 2002
  Related to Chapter 5

- **Authors**: Gouyon, F. Herrera, P. Cano, P.
  Title: Pulse-dependent analyses of percussive music

Conference: International Conference on Acoustics, Speech and Signal Processing
Year: 2002
Related to Chapter 5

- **Authors**: Herrera, P. Yeterian, A. Gouyon, F.
Title: Automatic classification of drum sounds: a comparison of feature selection methods and classification techniques
Conference: International Conference on Music and Artificial Intelligence
Year: 2002
Related to Chapter 3

- **Authors**: Cano, P. Kaltenbrunner, M. Gouyon, F. Batlle, E.
Title: On the use of Fastmap for audio information retrieval and browsing
Conference: International Conference on Music Information Retrieval
Year: 2002
Related to Chapter 5

- **Authors**: Zils, A. Pachet, F. Delerue, O. Gouyon, F.
Title: Automatic Extraction of Drum Tracks from Polyphonic Music Signals
Conference: International Conference on Web Delivering of Music
Year: 2002
Related to Chapter 3

- **Authors**: Gouyon, F. Herrera, P.
Title: Exploration of techniques for automatic labeling of audio drum tracks instruments
Conference: MOSART Workshop on Current Research Directions in Computer Music
Year: 2001
Related to Chapter 3

- **Authors**: Gouyon, F. Pachet, F. Delerue, O.
Title: On the use of zero-crossing rate for an application of classification of

percussive sounds

Conference: Conference on Digital Audio Effects

Year: 2000

Related to Chapter 3

- **Authors**: Pachet, F. Delerue, O. Gouyon, F.

  Title: Automatic Extraction of Rhythmic Structure From Music

  Conference: Sony Research Forum

  Year: 2000

  Related to Chapters 3 and 5

- **Authors**: Thornburg, H. Gouyon, F.

  Title: A Flexible Analysis/Synthesis Method for Transients

  Conference: International Computer Music Conference

  Year: 2000

  Related to Chapter 3

Appendix C

Sound distortion scripts

This appendix details the sound distortion scripts used in Section 2.3, on page 70.

System commands

- Resampling:
  ```
  $ sox wavfile.wav -r8000 soxedfile0.wav rate
  ```

- GSM encoding/decoding and upsampling:
  ```
  $ sox soxedfile0.wav soxedfile1.gsm
  $ sox soxedfile1.gsm -sw -r44100 soxedfile2.wav rate
  ```

- Filtering and volume adjustment:
  ```
  $ sox soxedfile2.wav soxedfile3.wav filter 500-2000
  $ sox soxedfile3.wav soxedfile4.wav vol 1.8
  ```

- Reverb application:
  ```
  $ sox soxedfile4.wav soxedfile5.wav reverb 1 2000 1000 700 750 760
  880
  ```

Matlab commands

- White noise addition:
  ```
  >> [x,fs,bits] = wavread( soxedfile5.wav);
  ```

```
>> SNR = 20;
>> Px = sum(sum(x.^2));
>> noise = rand(size(x))-.5;
>> Pnoise = sum(sum(noise.^2));
>> noisyX = x + noise*sqrt( (Px/Pnoise) * 10^(-SNR/10) );
>> wavwrite(noisyX, fs, bits, tempwavfile.wav);
```